Robot Awakening

The Time of Artificial Life

Robot Awakening

The Time of Artifical Life

By
SIMON KING

CONSCIOUS CARE PUBLISHING PTY LTD

ROBOT AWAKENING
The Time of Artificial Life

First Published 2018 by: Conscious Care Publishing Pty Ltd
PO Box 776, Rockingham, WA 6968, Australia
Phone: (61+) 1300 814 115   www.consciouscarepublishing.com

First Edition printed January 2018.

National Library of Australia Cataloguing-in-Publication entry:
Author: King, Simon 1950-
Robot Awakening / by Simon King
ISBN 780648085454 (Paperback)
Rocky Hudson, Editor.

Printed by Lightning Source
Typeset & cover design by Conscious Care Publishing Pty Ltd

ISBN: 978-0-6480854-5-4

Dedication

This book is dedicated to every child who has ever experienced the fascination of toy robots and dared to dream of real robots from the future. It is dedicated to avid science fiction fans worldwide who have been thrilled at the exploits of fictional robotic life forms from this planet and from far beyond our solar system. Most importantly, the book is dedicated to the inventors, researchers, designers, artisans, programmers, engineers and other scientists past and present who have pursued the goal of creating robots in many forms and configurations, and who continue to endeavour to progress such technological investigations into artificially intelligent beings. Such innovative and inspiring research and development has encouraged me to compile this book.

Preface

Automata, robots and androids have been associated with the human civilisation in various forms and configurations for centuries, mostly as fictional fragments of science and the arts, but occasionally as prototype machines. These machines may assist people in their working lives, may provide entertainment or may exist simply to stimulate the imagination for future possibilities. Whether these entities are industrial tools relieving workers of their more onerous labouring tasks or humanoid robots simulating our species, they are usually created to serve humanity.

Science fiction has always illustrated many variations on how such entities may arise, inclusive of their arrival even from other planets. Scientific fact has demonstrated that humans can create them satisfactorily using artificial intelligence ('AI'). By tracing the historical origins of robotic development and notable technological advancements, a diverse array of robotic designs will be examined; from popular children's robot toys of the past and various sci-fi artistic genres, to the sophisticated 'intelligent' commercial androids and robotic communication systems of the present.

Using a combination of science fiction stories moulded to contemporary worldwide achievements in cybernetics and AI development, the book considers the evolution of the rapidly growing robotic society in today's climate, and the various potential implications of this progress. As synthetic artificial life forms continue to develop and evolve, such entities could eventually generate an entirely different society of the future.

A new technological age of robot awakening, whereby they become self-replicating and autonomous, may possibly be just over the current scientific horizon. Be prepared for the ride.

Contents

List of Figures

Replication of the Species

Replicants are just like any other machine. They're either a benefit or a hazard.

Rick Deckard, *Blade Runner (1982)*

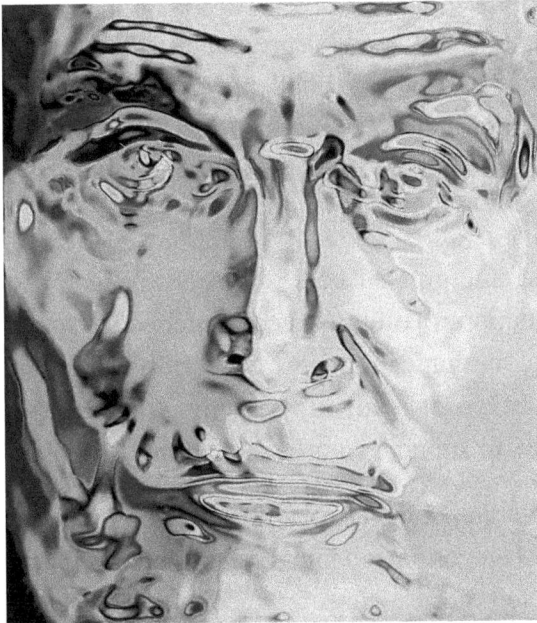

Figure 1: Replication Android (© Shutterstock)

ROBOT AWAKENING

Whichever way the matter is considered, robotic devices and machines, humanoid robots and artificially intelligent androids have been devised, designed and manufactured to serve humankind. Such creations initially amused and entranced the population in bygone ages, but ultimately have assisted civilisation in a vast array of fields. Most notably, they are being successfully used as dedicated social companions in health and in human care of the elderly, the chronically ill, and in rehabilitation of the disabled. Robotic technology has recently been impressively applied in the introduction of driverless automobiles and autonomous, gargantuan mining trucks and ore trains, with many more widespread industrial applications to follow.

The automation of industrial processes initially provided the essential grounding for robotics, particularly where the tasks were onerous, dangerous or excessively repetitive for people. However, their success at relieving humans of many arduous tasks has spread into far more domains requiring sophisticated scientific assistance. In the domestic market, an enormous range of useful household appliances are developing to ease the burdens of housekeeping. In education and leisure, functional 'smart' tools are providing end-users with new learning challenges. In medical processes, robotics are providing extraordinary advances in complex diagnostic analysis and surgical procedures, whilst remarkable use of innovative android technology is being utilised in space exploration ventures.

One of the fastest emerging success stories has been in providing a readily accessible information revolution in computerised systems and the routine mechanisms of communicating on the worldwide Internet. As a consequence, the human race is evidently becoming increasingly reliant on such artificial providers to support their lives, and this will only increase.

In the earliest developments of robot mobility, the robots were designed ostensibly in our image, as 'mechanical men' which possessed similar functionality. People felt more at ease in the company of beings that resembled us, even if they were constructed of metal. Future advancements to artificially intelligent synthetic life forms, such as human-like androids, has created robots that appear to act almost indistinguishably from us. Humankind has effectively commenced the first experimental steps to self-replication.

Eventually, sometime in the future, everybody will need to confront a society potentially dominated by its own robotic creations. *Homo sapiens* have always considered themselves to be the superior species on this planet, but as 'life forms' created of artificial intelligence develop and multiply in society, it may become apparent that we have new neighbours. Intellect and emotions are traits that humans possess, and are not qualities designed on a computer. However, a sophisticated and highly intelligent artificial species with dedicated logic and reasoning has the ability to learn, to absorb this learning, and most importantly, to morph.

There is also the interesting technological concept of such life forms communicating between themselves without any direct human intervention or control. This may include a centralised control where many robotic devices are directed by one artificial intelligence source. If it sounds like a combination of science fiction and a vivid fanciful imagination, such implausible concepts usually originated from such realms. Then again, even the concept of actual artificial intelligence was only really found in fictional magazines midway through the 20th century.

The widely acclaimed English poet, playwright and actor William Shakespeare provides the following wistful thoughts: 'There are more things in heaven or earth, Horatio, Than are dreamt of in your philosophy [learning].'[1]

History of Discovery

Hire the Artificially Intelligent.

Bumper sticker, *circa 1995*

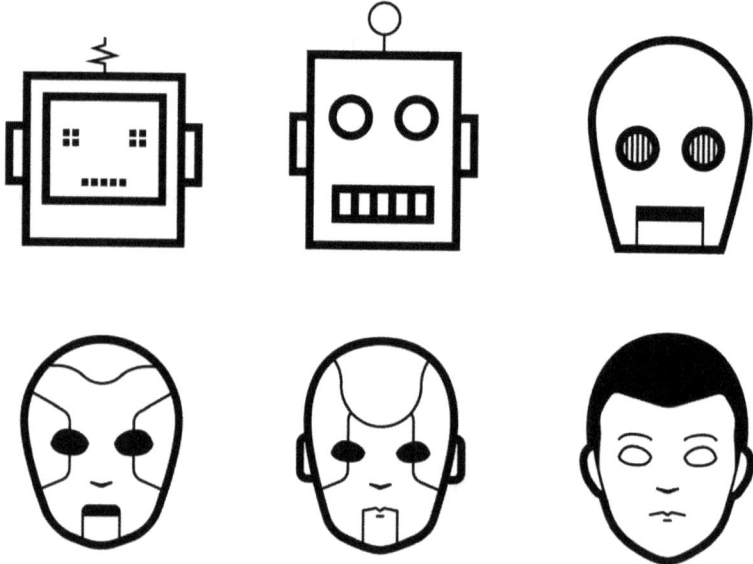

Figure 2: Artificial Intelligence Evolution

(© Shutterstock)

To understand the history of discovery of robots, it is pertinent to first clarify some essential terminology. An automaton by the earliest definition of a robot can mean a mechanical figure or contrivance constructed to act as if by its own motive power (1605-1615). The derivation is from Latin for 'automatic device' and Greek for 'acting without human agency'. This description also extends to both a person who acts in a monotonous, mechanical/machine-like way without active intelligence, and something capable of acting automatically or without an external motive force.[1]

'Automaton' first came into popular use in 16th century France after the term 'automate' was used to denote a machine with a self-contained principle of motion (1532 AD). In the 20th century, it encompassed more specialised terms including robots, androids and cyborgs.[2]

Automata are the mechanical models of living beings, operating either as robots or machines under their own power.

> 'Until the latter half of the 20th century, the value of actual real-life robots – or automata, as they had usually been called – lay almost solely in their ability to surprise and amuse the onlooker.'[3]

There are innumerable examples of popular and peculiar robot inventions, of automata capable of playing musical instruments, exhibiting feats of supreme strength, performing stunts or perhaps responding thoughtfully to intelligent questions from the public. The subject matter is probably a book in itself. Suffice to say that until the revolutionary introduction of industrial robots replaced the need for humans to perform many tedious, repetitive, arduous and sometimes unpleasant tasks, not much changed.

Such a transition, from treating automata as sophisticated toys to accepting their capabilities to conduct tasks better than humans, was paramount. Automata could provide a far superior quality of work by virtue of their precision, consistency and most importantly, their ability to operate continuously.[4]

By stark contrast, '... according to the Oxford Online Dictionary, the word "robot" comes from robota, the Czech word for 'forced labour' [servitude].

ROBOT AWAKENING

The term was coined in [Karel] Capek's 1920 play R.U.R. 'Rossum's Universal Robots'....'[5] It pertained to a 'synthetic worker' that had been artificially created from organic material – fictional automata.

To fully appreciate the history of robot discovery in the tradition of Western intellectual ideas, the explanatory journey must first embrace the earliest automata of human civilisation. The Egyptians were probably the first to build mechanisms to imitate parts of the human body, when they attached mechanical arms to the statues of their gods. These arms were operated by priests who claimed to be acting under inspiration from the gods.[6]

The great Greek engineer Ktesibios/Ctesibius (circa 270 BC) was responsible for applying a knowledge of pneumatics and hydraulics to produce the first water-powered organ, water clocks with pneumatic moving birds to sound the hours, as well as other automata.[7] From the 3rd century BC, automaton-making by the ancient Greeks flourished. Figures of humans, animals and mythological creatures were used to depict various scenes and to entertain viewers. Such ingenious designs were predominantly to display various mechanical principles at work.[8]

Over the 15th-16th centuries, European engineers and artisans continued to refine and improve the complexity of the machines they built, with some automata strongly resembling humans in appearance and motion, inclusive of possessing proper expression and sentiment.[9] Automatons resembling human figures and propelled by hidden mechanisms were used in churches to create an illusion of self-motion to the parishioners.[10]

One of the first recorded detailed designs of a humanoid robot was made by Leonardo da Vinci around 1495 AD, resulting from his studies of human anatomy. His mechanical man or 'anthrobot' mimicked the appearance and the movements of a Germanic knight in a suit of armour. The design indicated it was able to sit up, wave its arms and move its head via a flexible neck while opening and closing its jaw.[11] It relied upon a complex system of cables and pulleys which articulated the robot's parts with anatomical accuracy. His project certainly displayed his incredible endeavours to attain a real robot.

Figure 3: Musical Automaton

(© Shutterstock)

In the late 16th and 17th centuries, a renewed interest in understanding the human body and the cosmos in mechanical terms fuelled ongoing development of automata.[12] By the beginning of the 18th century, many automata were being built, some of which could act, draw, fly, or play music. Even miniature automatons resembling humans or small animals became popular as expensive toys. These were intricate mechanised 'puppets' imitating the actions of living beings. By the 18th century, the Swiss had the skilled craftsmen and technology (due to an established clockwork and watchmaking industry) to develop the first recognisable and responsive human-like device of reprogrammable precision machinery.

The world's first 'robot' was created in the 19th century by Zadoc Dederick and Isaac Grass in the USA (1867-68 AD) and was known as the 'Newark Steam Man'. It would be the first of a generation of steam-powered walking humanoids between automata and cybernetic machines.[13] This is '…a robot in a top hat with an enamel face, designed to pull carriages and to look as human and un-frightening as possible… Steam is generated in his trunk [three horsepower engine] that drives the mechanism which moves his limbs'.[14] However, it would be the onset of the 20th century that would accelerate robot development markedly.

The first British *robot* was built in 1928 by First World War veteran Captain William Richards and aircraft engineer Alan Reffell. It was named Eric the mechanical man and resembled a knight-in-armour. On suitable commands, '… he was able to stand up, bow, look to his left and right, sit down and with appropriate gestures, deliver a four-minute opening address'. He had some fascinating features including '… 35,000 volts of electricity causing blue sparks to shoot from his jagged teeth … inside his body … approximately 4.5 kilometres of wires'.[15]

A fitting introduction to this new invention may well be taken from the following partial verse by Louis Bertrand Shalako in his poem *Mr Robot* first published in 2010:

> … Here I come
> Cryogenic heart, skin a polished silver
> One thing I am glad of
> For this I thank my builder
> I can never rust…[16]

Between 1937 and 1938, Westinghouse Electric Corporation in the USA built the impressive ELEKTRO. It was a 2.1 metre humanoid steel robot with an aluminium skin that could walk, talk, smoke, and respond to voice commands; it had a recorded vocabulary of more than 700 words. He was '… tall, chunky, gleaming, square-headed and actually quite fearsome-looking – like a sci-fi robot come to life'. [17]

In 1954, prolific American inventor George Devol devised the world's first truly programmable robot, which was an industrial robotic arm used in manufacturing assembly. It was named Unimate and was designated for

'Universal Automation'.

*****

It is perhaps the insatiable thirst for perfection that so motivates humans to succeed at robot discovery and development, and what better way to track such progress than through the magic of cinema. Depictions of fictional robots throughout cinematic history are a very creative way of acknowledging their development and there have been many notable examples to recall. As early as 1908, in the film *The Fairylogue and Radio Plays,* there was Tik-Tok that '... runs on clockwork springs which need to be wound up, and is thought to be the first "mechanical man" in 20th century literature'.[18]

From the early 1930s to the mid-1950s, such creations were depicted purely as a mechanical figure or sheet-metal robot in films such as *The Phantom Empire* (1935) and *Wizard of Oz* (1939). Thereafter, they were '... as a complex humanoid robot with an electronic brain, linked to the various "positronic" robots... *Forbidden Planet* (1956)'. In the modern era, there were androids and cyborgs in films such as *Blade Runner* (1982) and *Aliens* (1986).[19]

*****

Perhaps one of the most prophetic sci-fi robot movies of the modern era has been the cult classic *Hardware* (1990) by British Satellite Broadcasting/Palace Pictures, in which the robotic head of a defunct military cyborg takes on a life of its own. The Mark 13 cyborg had been dismantled and discarded due to a major design fault, but its head was salvaged from the wastelands following a nuclear war. People have five senses but this cyborg appeared to only have two – the ability to kill and the unrepentant resolve to survive by self-repairing. Together, this makes for a lethal combination.

The cyborg reactivates after being given as a gift to an unsuspecting human and proceeds to rebuild its entire body. With a determination to exterminate anybody in its way, the Mark 13 inflicts a tremendous toll on humans in its vicinity by using an arsenal of high-tech weaponry. In the final fatalistic scenes of the film, the question of whether such an indestructible cyborg can actually be stopped is resolved in fitting dramatic fashion. It provides an entirely new meaning to 'taking a shower'. May you rust in peace, Mr Robot.

'It is hard for a robot to wipe mud from its eyes when it has whirring buzz saws for hands.' [20]

*****

So in the modern era, what robotic discoveries and advancements appear almost without parallel? History will probably verify that the rapid development of computers through the 1960s-1970s, propelled to an extent by the scientific need for Space Age technology, signalled the dawn of this era. From a diversity of experimental computer-controlled walking machines to various industrial robots, the ongoing evolution of programming continued to yield results. Artificial intelligence research was underway. The race was on to develop sophisticated anthrobots capable of truly simulating human behaviour.

For a suitable example, I select the versatile WABOT-1 (WAseda roBOT) invented by the highly esteemed late Professor Ichiro Kato (circa 1970 ~1973) at Tokyo's Waseda University in Japan. This hydraulically powered, two legged walking 'android' with two arms and hands was the first full-scale anthromorphic (humanoid) intelligent robot in the world. It had systems for controlling limbs, vision and conversation, including the abilities to recognise and handle objects. Disproportionately large feet designed for stability effectively confined it to shuffling.[21] It was also estimated that robot had the mental faculty of an 18-month-old child.[22] One ponders if this attribute extended to the usual tantrums and raucous behaviour exhibited by such youngsters.

Figure 4: WABOT (WAseda roBOT)-1 (1973):
(© Humanoid Robotics Institute, Waseda University, Tokyo, Japan)

*****

The development of a self-regulating, bipedal, walking humanoid robot that could also successfully ascend and descend stairs independently was another significant advancement for its time. Early prototype robots invented by Honda engineers in Japan commencing in the mid-1980s eventually lead to the P-series humanoids, and the impressive P2 robot in 1996. This innovative development and its refined successor the P3 one year later, represented a remarkable transition in achieving simulated human mobility.[23]

ROBOT AWAKENING

The success of a humanoid robot capable of performing many difficult mobility tasks by regulating its movements to suit each specific hazard encountered came with a minor drawback. The P3 was 1.6 metres in height and weighed a massive 130 kilograms, perhaps resembling a near-human size too closely. Further evolutionary change ensued and in 2000, ASIMO (Advanced Step in Innovative Mobility) was created, weighing a modest 54 kilograms and standing a mere 1.20 metres tall.

The ASIMO suite of fully autonomous humanoid robots developed through the 2000s achieved a remarkably broad range of intelligent capabilities. These included enhanced abilities to interact with humans by interpreting Importantly, R2 as Robonaut is known, is programmed for close contact, making it the optimum user-friendly assistant in confined working areas. body postures and gestures, facial expressions and voice commands, plus the movements of multiple objects. ASIMO could make informed decisions and act autonomously by integrating relevant information about the immediate surrounds, and operate harmoniously within a real-life human environment.

Later generations of ASIMOs can work collaboratively with each other to achieve the most efficient outcome for their tasks.[24] The little humanoid robot that resembles a small astronaut carrying a backpack certainly is versatile. It can even run continuously at the incredible pace of 9 kilometres/hour.[25] These certainly are innovative and impressive advancements in modern robotics.

Figure 5: ASIMO and P3 Humanoid Robots

(© Honda Motor Co. Ltd., 1997-2011)

*****

In February 2011, the National Aeronautics and Space Agency (NASA) launched its space shuttle Discovery on a unique journey to the International Space Station (ISS) orbiting our planet. On board was Robonaut 2; the first humanoid robot deployed to become a permanent resident of the ISS to assist human astronauts. The torso-only robot prototype was subsequently enhanced with two legs in 2014 to provide manoeuvrability inside the confined quarters.[26]

Robonaut was jointly developed by NASA and General Motors to work like its human counterparts using the incredible dexterity and sensitivity provided by equivalent hands and arms. It is the perfect robotic assistant for every astronaut and has no need to eat, sleep or worry about the long hours in space. Ideal for tasks likely to be too difficult or particularly hazardous for the human crew.

ROBOT AWAKENING

Importantly, R2 as Robonaut is known, is programmed for close contact, making it the optimum user-friendly assistant in confined working areas.

*****

As the technological world embraces knowledge sharing between multiple robots and increasingly smarter 'bots', fuelled by an insatiable need for personal server robots to assist humans, the race is on to design that perfect companion or industrial worker.

Imagine a future world that has eliminated almost every repetitive, tedious and arduous task by the introduction of a gargantuan robot empire servicing the human population. A world where science fiction has materialised into science fact in so many ways. A world where generation after generation of advanced self-replicating computers and sophisticated androids proliferate. The only remaining challenge for us, then, may possibly be in seeking other planets to colonise, unless it is already too late.

Figure 6: All this will belong to you one day Junior

(© Shutterstock)

As an interesting perspective on such potential technological challenges likely to face humankind, I provide one possible viewpoint about the introduction of server 'intelligent' androids.

There was a service station that simply could not make a profit from selling fuel to passing motorists, despite offering first class service. The solitary human attendant would diligently check each vehicle's engine oil, cautiously inspect the water level of the radiator, thoroughly test the air pressure of each tyre, clean the windscreen and dispense fuel. Afterwards, the hard-working friendly attendant would operate the cash register to finalise each transaction, and often sold extra merchandise to the motorists, like hot food, confectionary or magazines. Frustrated by the dire financial situation, the owner of the service station replaced the attendant with four androids resembling humans to tackle the same mundane tasks.

For a period, everything worked perfectly and profits started to increase. Motorists commended the owner on the fantastic efficient service and the extreme politeness of each android. Nothing was ever too difficult for these robotic servers, they never complained, took any time off work or made any mistakes and most importantly, worked for free. The correct tyre pressures were always maintained, windscreens were kept spotless, the exact change was always carefully returned from the cash register and profits increased.

Then one day, the androids were nowhere to be seen at the service station, and the same solitary attendant was back doing the same work. A bemused motorist stopped and enquired about why the androids were no longer at work. The attendant's expression was grim and serious as he explained that they had 'misbehaved'. On one particularly hot summer day, instead of checking the radiator water and filling the fuel tank, they poured petrol into the radiator and filled the vehicle's tank with water. Windscreens were left uncleaned and motorists were chased by the androids intent on checking their personal air pressures. Money was retrieved from the cash register and thrown at each motorist. Complete pandemonium was evident whenever a vehicle pulled into the service station.

The motorist then enquired what had caused this incredible behaviour. Had their programming malfunctioned? Was a computer virus at fault? Perhaps there was some other technological glitch? The human attendant then stared the motorist straight in the eye and quietly replied 'Nothing as complicated as those reasons. It was simply a matter of too much Artificial Intelligence', and he went back to work with a broad smile.

Science Fiction

Science fiction deals with improbable possibilities, fantasy with plausible impossibilities.[1]

Miriam Allen de Ford, *Elsewhere, Elsewhen, Elsehow (1971)*

One of the most puzzling questions to answer with any certainty is whether robots, humanoid androids and devices capable of exhibiting artificial intelligence originally evolved from the fanciful imagination of science fiction or from the inquisitive motivation of mankind to develop improved ways to live on this planet and elsewhere. Scientific and technological advances often have a sound basis in imaginative thinking, so it would be reasonable to assume that the fictional realms of science played some significant role in some of these futuristic creations.

If I was to look more closely at robots and their derivatives, it would be apparent that there is some commonality of attributes shared between them. Rationality, objectivity and unambiguous problem-solving logic skills are the most obvious of the human traits endowed upon them. Conflicting senses such as feelings or emotions requiring a social sense of order have sensibly not been allocated in order to preclude confusion and disorientation. These human traits would surely require an extreme shift in thinking in order to create individual robots capable of such self-thought.

In some ways, this is man's ultimate shutdown safety valve, designed to prevent the invention from making the critical transition to become human in far more than simply physical appearance. Having been a science fiction fan all my life, it is apparent that underlying the considerable quantity literature and film produced about the robotic world is the dread of a robot revolt. The result is almost always the slave uprising that results in the eventual enslavement and potential destruction of humanity.

*****

One of my favourite science fiction episodes of *Doctor Who* (the English time-traveller in the BBC series) is the ominously titled *The Robots of Death*, first screened on television in early 1977. It was set inside a large futuristic sand-crawling mining vessel extracting valuable minerals on the surface of a desert planet and operated largely by humanoid robots. The small technical crew of humans managing the miner are slowly and mysteriously eliminated by robots in contravention of the First Law of Robotics (robots shall not harm humans), until there are scarce survivors.

Although 'the revolt' involved reprogramming of select robots to kill all humans, the mastermind scientist behind the plot had actually been raised by robots. In the final harrowing scenes when all appears lost, one of the robots assisted in the rescue by self-destructing and thus stopping the super-robot directing the revolt. In effect, it prevented further harm to humans at the cost of its own existence.[2] Even science fiction tales need to follow the Laws of Robotics.

*****

A far more subtle revolt of androids was convincingly portrayed to deadly effect in the 1973 cult classic movie *Westworld* produced by Metro-Goldwyn-Mayer. It was set in a futuristic high-tech theme park open to holidaymakers seeking to fulfil their fantasies. The park provided realistic human-like androids to interact with the tourists, comply with their requests and simulate actual life in three distinctly different worlds, namely the Roman Empire, the Medieval period, and the Old West or *Westworld*. For the unfortunate guests savouring life in these worlds, a major computer malfunction progressively resulted in a totally unplanned experience.

The android characters systematically turned on their human guests, with somewhat gruesome results. Despite their programming, the androids commenced wreaking havoc in various ways, ultimately eliminating their holiday guests and the park staff with only one rare exception.

Figure 7: Robot Stare
(© Shutterstock)

In the Old West, one particular android known as the Gunslinger – played magnificently by the late American actor Yul Brynner – epitomised the intent behind this carnage. This android was programmed to start gunfight duels with guests and always lose. With the robot revolt underway, the Gunslinger no longer lost any gunfights and despite many attempts to destroy it, relentlessly pursued the only surviving human.

The Gunslinger has even been rated in the world's top five deadliest fictional robots and cyborgs, sharing this accomplishment with the Daleks and the

Terminator amongst the nastiest of the bad and mad metal machines. Considered as dark, impassive and single-minded, the Gunslinger proved that there is nothing so terrifying as efficiency, and this killing machine even kept going after he was half-blinded [by liquid acid] and set on fire.[3] Have we got a vacation for you!

*****

In order to realistically examine the realms of robotic fiction, who better to discuss than Isaac Asimov, the prolific and acknowledged doyen of contemporary science fiction literature who '…wrote nearly 40 robot stories, so many that they could not be reprinted in a single volume'. Of these, there were stories that '…added significantly to the intellectual and emotional consideration of the robot that Asimov began in 1939'.[4]

Of most interest to me about Asimov's works are his insightful and famous Three Laws of Robotics, as well as his conceptual vision of a futuristic world with a dedicated robot society. These laws would appear for the first time in a short story (the third robot story written by Aismov) entitled *'Liar!'* that was published in the *Astounding Science Fiction* magazine in May, 1941.[5] So much has been written about these laws in innumerable books concerning robots that no further discussion is needed here, although I think that they represent important foundation blocks for human understanding of robotics on this world.

The Three Laws of Robotics (Asimov's Laws):

> 1: A robot may not injure a human being or, through inaction, allow a human being to come to harm;
>
> 2: A robot must obey the orders given it by human beings except where such orders would conflict with the First Law; and
>
> 3: A robot must protect its own existence as long as such protection does not conflict with the First or Second Law.[6]

Admittedly, these laws are fictional, yet why would a robot that is capable of reasoning, learning and self-improvement willingly elect to harm a human being? The obvious response is for its own preservation, in order to prevent any unauthorised shutdown or disassembly. Assume that for just a moment, however, there are more potentially surreptitious reasons. This may be maleficence/revenge or hostility, including jealousy about their human counterparts, as indicated in the following humorous quotation from robotics expert Daniel H. Wilson:

> 'Robots have no emotions. Sensing your fear could make a robot jealous and send it into an angry [white hot robot] rage.'[7]

Asimov provided variations on a theme when he wrote about our planet in the future being overpopulated '... a society heavily overweighted in favour of humanity, with the robots unwelcome intruders' (*The Caves of Steel*, 1953), or conversely, a fictional planet overrun by robots '... an almost pure robot society with only a thin leaven [pervasive influence] of humanity holding it together...' (*The Naked Sun*, 1956).[8]

The contrast between the two cultures described in these stories is significant. With Earth's homogenous society choosing to live in massive cities that efficiently fulfil most of their essential necessities and thus avoid the wide open spaces, robots serve to provide all agricultural and mining use of the open country. The consequence is that Earth becomes concerned with competition from robots and the threat that they may eventually replace people as well. Alternatively, on the relatively unpopulated fictional planet, where there are only twenty thousand people and two hundred million robots, and wide open estates up to ten thousand square miles, people seldom come into personal contact due to the vast distances involved, and as a result, avoid other human beings.[9] Robots virtually control almost every aspect of life and people barely matter. With such vastly contrasting fortunes for people, depending upon the extent of robot society, I am certainly glad these stories remain fictional, albeit rather sobering.

*****

For those contemplating total automation of their factory, the following story from the immensely popular original American television series *The Twilight Zone* hosted by Rod Serling from 1959 to 1964 suggests some very careful thought.

ROBOT AWAKENING

In a science fiction episode entitled *The Brain Centre at Whipple's* aired on CBS (Columbia Broadcast System) in the last season of the series, Mr. Whipple, the owner of a vast manufacturing corporation, decided to significantly improve factory efficiency and output by installing a totally automated machine called X109B14. Most workers became redundant, and despite the protestations of some that the value of living human beings far outweighed that of machines, Mr. Whipple continued to be obsessed with installing more machines to completely automate the factory, and as a result, the company's board of directors retired him.

Figure 8: Android Factory Workers

(© Shutterstock)

His replacement was a humanoid robot which had Mr Whipple's quirky mannerism of twirling his watch. The ominous closing warning coldly narrated by Rod Serling at the episode's end said it all: '… Man becomes clever instead of wise; he becomes inventive and not thoughtful; and sometimes, as in the case of Mr.Whipple, he can create himself right out of existence, as in tonight's tale of oddness and obsolescence, in the Twilight Zone'.[10]

*****

What if this historical battle between the brain of a human and the product of the human's brain has already been decided? Take a fully robotic hotel of the potential future, for example, where everything is controlled by robots of various persuasions using voice activation and sensory movement technology. The only humans involved are the guests. Upon arrival at the hotel's entrance, guests' luggage is collected and transported by an automaton mobile carrier along an access ramp to reception, passing through movement-activated doors. There are no robot-unfriendly staircases or steps in this futuristic hotel.

Guest registrations are handled via an android concierge which re-directs the luggage to the automatic lift and then to the necessary floor level. Every room/suite has an assigned small robotic ('cute') valet to welcome each guest and provide personal assistance as required during their stay. It provides an entirely new meaning to the phrase 'room service'. Every facility inside the accommodation is activated by specific voice recognition of the guest and operates to maximum efficiency. Electrical appliances, bathroom and kitchenette taps and room lighting operate to whatever standard the guest verbally requests and there is no need to complain. However, there is one drawback. Does this hotel also accept android (non-human) guests? Better look over your shoulder to see if your neighbour is actually human.

Industrial Bots

... you just can't differentiate between a robot and the very best of humans.

Isaac Asimov, *I,Robot (1950)*

Industrial robots ('Bots') are the foundation for robotics on our planet. If the task at hand demands tedious repetition, monotonous diligence, tough heavy-duty effort or perhaps difficult and relentless painstaking concentration, who better to perform the work than an artificial robotic device. Certainly in most cases, humans probably do not actively seek such mind-numbing repetitive work. This is where the only realistic challenges lie in ensuring there are neither production mistakes made with products handled on complex assembly lines, nor personal injuries incurred, particularly when handling extremely hazardous materials. Heavy lifting or awkward manoeuvre tasks potentially likely to injure workers are always destined to be eventually handled by machinery, as far as reasonably practicable. If the work is dirty or unpleasant, humans rarely opt for it.

This reminds me of a subtle witticism penned by the famous and highly talented late British cartoonist John Morris where two employees of an automobile assembly plant are gazing at the factory noticeboard that nominates the employee of the month, and one of workers blandly responds by stating that 'the robot won it again!'[1]

In its simplest terms, the human body is organic and comprised of 'soft flesh, fragile organs and brittle bone' when compared to metal robotic machinery for example, and thus is particularly sensitive to environmental hazards. Industrial bots also provide a substantial range of attractive qualities suitable for the mundane or repetitive work functions; they can provide dexterity and precision, are methodical with minimal deviation from the task, and most importantly, they rarely stop and so can operate around the clock if required. This makes them ideal for product reproducibility and quality control inspections. Bots are also safe and do not object to working long hours. In some organisations, they are affectionately known as 'an additional veteran worker' that quietly goes about its business each day.

So what are 'industrial robots'? The precise definition of an industrial bot is somewhat technological: 'A reprogrammable, multifunctional manipulator designed to move material, parts, tools, or specialised devices through variable programmed motions for the performance of a variety of tasks.'[2] An industrial bot could be as simple as a bottle capping machine whose electronic eyes perceive and interpret complex visual patterns, or a sequential assembler machine moving a part from one place to another, or something as complex as a reprogrammable articulated arm for spot welding or riveting.

Of course, this definition could be expanded considerably depending upon where the bots operate in the world. Japan has always been the world's leading user of robots since the 1970s, with a stringent focus on automating factories, providing efficient alternatives to smaller businesses, and investigating robotic opportunities in the wider community. Its acceptance of bots was so pronounced that Japan successfully went from an importer of robot technology to an exporter in just ten years.[3]

Another more current statistic about this penetration of robotics in Japan's factories is that of robot density. In 1980, Japan had 10 industrial robots per 10,000 manufacturing employees, but by 2010, it had more than 300 per 10,000 employees. There are now in existence fully automated factories where humans are not required on site in various parts of the world, including Japan.[4]

Figure 9: Bottle Capping Bots

Consequently, I provide the following distinctive categories of such industrial bots based upon Japan's technological use:

- Manually manipulated by a worker;
- Predetermined/fixed time sequence robot for position and operation;
- Variable time sequence robot where movement can be easily changed;
- Playback robot where its operation is programmed manually and then memorised by the robot which can repeat the movement on its own;
- Numerical control robot that is programmed by numerical coded data; and
- Intelligent robot that determines its movement through information obtained from its own sensory ability.[5]

It is also important to realise that many industrial bots are not 'intelligent' bots. They may represent production machinery that perform repetitive, often standardised, functions which could also be better undertaken by humans or even non-robotic production devices. The fundamental difference is that such bots are ruthlessly efficient. Most are incapable of making independent choices in their processes unless so programmed, but they still fulfil an essential economic role within industry. I particularly like the following subtle account of such bots in a factory where it was feared by managers that their introduction would distract the workforce:

> '...workers would be distracted by the music the robots [computerised carriers] play to warn workers of their approach. This has not been a problem, however...In fact, I've noticed some of the employees humming along with the robots as they work.'[6]

Small business operators, however, may perceive bots somewhat differently:

> 'If you make a mistake and choose the wrong robot, it will sit there in the shop. Programming is very important. If you can't make your robot do what you want it to, it will just push you around.'[7]

So how are these industrial bots distinguishable from other robots? Much of this answer lies in their appearance and in their performance capabilities. Unlike various other robots (such as personal service units and intelligent androids, for example), industrial bots only operate in a highly structured environment performing prescribed tasks, and are often segregated from humans. Only trained people interact with them. Their configurations in factories vary markedly depending upon the particular industry, from automobile manufacturing and assembly works to sterile clinical environments and beyond.

Their design is engineered to efficiently deliver a prescribed service such as welding, machining or painting, and consequently they appear in a myriad of layouts, such as a single or multiple flexible arm, or possibly as a mobile robotic transporter machine. The important question is: do they occasionally malfunction and commence actions not programmed by humans? Do they ever break Asimov's First Law of Robotics (1950) – that is, 'A robot may not injure a human being or, through inaction, allow a human being to come to harm?'[8]

If they do, is it possible that an industrial bot has operated as an 'intelligent' robot and made a pre-determined decision to harm, or am I being too fanciful? Only an intelligent robot can make informed decisions known as performance choices contingent upon sensory inputs (its own sensory inputs). If an industrial bot did 'over-ride' its programming in some unique manner, would we truly know what caused it?

The definition of bots provided by the Japan Industrial Robot Association in the 1980s might suggest otherwise: 'A machine capable of performing versatile movements resembling those of the upper limbs of a human being or having sensory and recognition capacity and being capable of controlling its own behaviour.'[9] The last part of this definition provides an interesting foresight.

From my pragmatic perspective, I consider such bots as simply the drones of robotics rather than any higher level of robot. Notwithstanding, the question remains unanswered and to this end, I revert to American horror fiction author H.P.Lovecraft's warning:

> 'Children will always be afraid of the dark, and men with minds
> sensitive to hereditary impulse will always tremble at the thought
> of the hidden and fathomless worlds of strange life …'[10]

Consider a fully automated assembly factory that operates continuously without human intervention other than for maintenance and quality inspection purposes. Pre-suppose that the factory operates with a combination of drone bots for the mundane repetitive functions and intelligent 'smart' controller robots for the sophisticated requirements to manipulate and assemble parts despite any unplanned eventualities. Back-up robots are also available to replace defective robotic units.

Now consider that every so often, a solitary individual part is assembled slightly out of synchronisation to its design, yet still passes through the entire system undetected despite the intense scrutiny of various robotic devices. A simple obvious example would be an uncapped bottle of soft drink that missed its turn on the assembly line. A complex example would be a critical connecting linkage pin in an automobile's braking system that was installed incorrectly, yet the vehicle's brakes still function correctly for quite a period.

Several months later in the right circumstances on perhaps a wet or icy winter's night, the linkage is severed and the automobile crashes. The primary cause of this event may be undetected unless it is a common occurrence elsewhere in the world in products originating from the same factory. If it is a random event or involves different critical parts from the same factory, it becomes virtually impossible to trace or is assigned as a quality control issue at the time.

Of course there are people who would sensibly argue that every process system is subject to random aberrations from time to time, based upon scientific probabilities. Defective or misaligned parts may still pass through the quality system undetected on rare occasions, however unlikely or minute, without some form of human intervention. Programming safeguards also presumably minimise this problem or do they? The point that I raise is whether from time to time such defects are inherent in robotic technologies, as such machines do not have any sensitivity to such consequences – do they really care if the part is defective and that potential calamitous outcomes for humans may eventually result?

As organisations rush to automate their factories and release people from the many mind-numbing and dangerous tasks involved in production assemblies, the prospect that reliance upon robots to deliver the same, if not a superior and more efficient, service needs to be kept in the right perspective – after all, they are only machines without feelings. They might have sensors for sight, touch and pattern recognition, but not yet intellect.

Robot or Android

**Even I can't bear to see or touch myself.
I, who was once, once comely,
who was always a lover of beauty.
And now I have to live in this exile.
I have to live amongst androids
because androids do not see as we see.**

Sharaz Jek, *The Caves of Androzani (1984)*

What is the difference between a robot and an android? Robots are immensely diverse in both their appearance and behaviour. The physical appearance of a robot depends largely on its function – when it looks like a machine, it is a 'mechanoid' and its efficiency will be suited to the task(s) for which it is designed. If resembling a human or an animal, it is called a 'humanoid'. A humanoid robot has body parts that resemble those of the human or animal body, such as heads designed to replicate human facial features for ease of social interaction.

Humanoid robots that are built to aesthetically resemble humans are 'androids' and are designed to simulate human performance – the characteristics and behaviours of a human. This would assist them when interacting with people or when operating in environments optimised for human use.[1] It may also be a synthetic organism designed to imitate a human.[2]

An android is very different from that type of humanoid robot where similarity to humans is not considered fundamental. It has recently been characterised in functional terms as "an artificial system designed with the ultimate goal of being indistinguishable from humans in its external appearance and behaviour".[3]

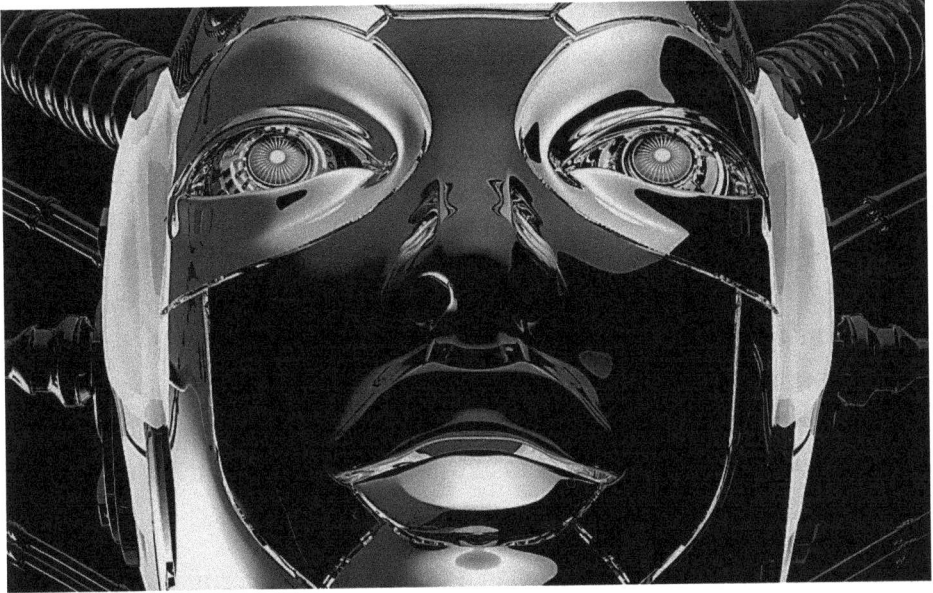

Figure 10: Robot or Android
(© Shutterstock)

*****

The first scientifically significant robots to be built in the world were in 1948/1949 and were named 'tortoises' by their creator British neurophysiologist Grey Walter, because of their appearance. Their development was important because these devices were created through a combination of ideas blended from engineering, mathematics and brain sciences. This application of ideas is known as 'cybernetics': the science of communications and automatic control systems in both machines and living things.[4]

The process was pioneered in 1948 by mathematician Norbet Weiner who published *Cybernetics,* or Control and *Communication in the Animal and the Machine*, that sought to unify understanding of how information is processed in animal brains and the resultant intelligent behaviour so that it might be adapted and regulated to control machines.[5]

In its simplest terms, there are machines built to act deliberately and display characteristics usually associated with living systems, such as 'intelligent' robots attempting to imitate various aspects of life.[6]

*****

There is possibly no better example of an android story then the *Doctor Who* television serial *'The Caves of Androzani'* screened by the British Broadcasting Corporation in March 1984.[7] In 2009, fans of *Doctor Who Magazine* voted this epic serial as the best *Doctor Who* show in history. More than 6,700 of them ranked all 200 episodes produced since 1963, and this serial narrowly beat the hugely popular *Blink* (2007) featuring the stature-like weeping angels.[8] More recently, in a 2014 poll by readers of the *Doctor Who Magazine*, it was ranked the fourth best of all time amongst 241 *Doctor Who* stories.[9]

Androzani Minor is bleak desert planet riddled with subterranean caves that are frequented by bats which produce a compound unique for its life-extending capabilities. This is the only known source of such a drug and the inhabitants of the neighbouring planet have a high demand for it. Mining has been severely disrupted by an army of faceless androids directed by a mysterious masked renegade known as Sharaz Jek. Soldiers sent to quell the uprising face insurmountable difficulties in successfully defeating androids in a complex system of caves also inhabited by a rather fearsome and lethal dragon-creature native to the planet.

Jek's subtle manipulation of the war relies to an extent upon his abilities to substitute realistic androids for people at various times, most notably the second-in-command of the soldiers known as Salateen. His clever deception of using near-perfect humanoid robots to confuse and undermine his foe is a tangible reminder of just how difficult it can be distinguishing the real person:

> *Salateen:* I've been held prisoner at his camp for months now, sir, and what you thought was me is in fact an android, a spy for Sharaz Jek.

The General: You mean I've had an android as my ADC [Aide-de-Camp] without knowing it? … What a blind fool I have been.

The process also works in both directions:

Jek: How is it that you were able to walk past my androids?
Dr Who: I don't know. Maybe they just liked my face?[10]

Jek's mask hides his hideous facial disfigurement of melted flesh, the prime reason for using his android army to execute revenge upon towards those responsible for destroying his previous privileged life. He had initially built the android workforce to collect and refine the unique compound for profit, but was thwarted by his business partner who was responsible for him almost perishing in a scalding underground flood of boiling mud.

*****

For an excellent example of a humanoid robot dog, the popular metal mutt K-9 from the early *Doctor Who* series (1977 -81) and subsequent enduring television appearances in *The Sarah Jane Adventures* until November 2010, is the ideal selection. Not only is this loyal companion of the Doctor highly intelligent and resourceful, it is a survivor. K-9 suffered innumerable mishaps along the way, from depleted power-packs at very inconvenient times to becoming stranded by inhospitable terrain or being short-circuited, kicked, decapitated and blown-up.[11]

As a result of these encounters, K-9 has been reconstituted or replaced by newer versions a few times, ultimately finishing as the Mark IV version. Remarkably, this resilient little robotic pooch usually survives the most ferocious attacks by adopting a clever canine manoeuvre or dogged determination to overcome a seemingly invincible alien foe. Without the four legs of a normal canine, K-9 relies upon concealed wheels for propulsion and so lacks speed and agility. The powerful and deadly laser weapon within its nose and a vast computer intelligence more than compensate for those mobility deficiencies. It even has a cute robotic tail.[12]

The best way to describe K-9 is by the following sentiments:

ROBOT AWAKENING

Doctor Who, **The Warrior's Gate, Part 2 (1981)**

K-9: The accuracy of this unit has deteriorated below zero utility.

Adric: You mean you're worse than useless.

K-9: Affirmative.

Doctor Who, **The Armageddon Factor, Part 1 (1979)**

Dr Who: Where's your joy in life? Where's your optimism?
Romana: It opted out.

K-9: Optimism: belief that everything will work out. Irrational, bordering on insane.

*****

An essential feature of a realistic android must be its hands and fingers as they are crucial elements for any human. The humanoid Dexterous Hand built by the Shadow Robot Company circa 2010 is certainly just such a commercially-available marvel. Its remarkable approximation to the human hand includes 24 different joint movements through the fingers, thumb and the wrist to simulate a range of movement equivalent to that of a typical human being. Finger tips are also fitted with highly sensitive touch sensors for that extra sense of realism. Its design selects 'the best possible initial grasp' automatically, either from a database of previously learned objects or by further computed assessment to ensure the most stable grip is executed. Once the object is carefully gripped, a further more stable grasp is safely applied autonomously. Customised versions are also manufactured to suit specific applications.[13]

Given the variety of handshake greetings that are possible between humans; from finger-crushing grips to soft, weak grasps that are barely detectable, it is intriguing to know how a like-fitted android would adjust its grasp in sociable settings rather than during its industrial functions. It would give an entirely new meaning to the most appropriate handshake.

*****

For a creation certainly not an android but more akin to a mechanical man from the late 1950s, *Cygan* the robot (also known as 'Gygan') was unusual by most standards of the times. Invented by the Italian engineer and aero-modeller Dr Piero Fiorito in 1957, it stood a gigantic 2.5 metres in height, weighed almost a massive 500 kilograms and comprised 300,000 parts.

The advanced technological design was kept relatively simple yet effective and was operated remotely by radio control. *Cygan* was particularly versatile in its movements; walking forwards or backwards, turning, raising or lowering each arm singly or together, and moving or crushing objects upon request. It even accepted spoken commands and responded to light signals.[14]

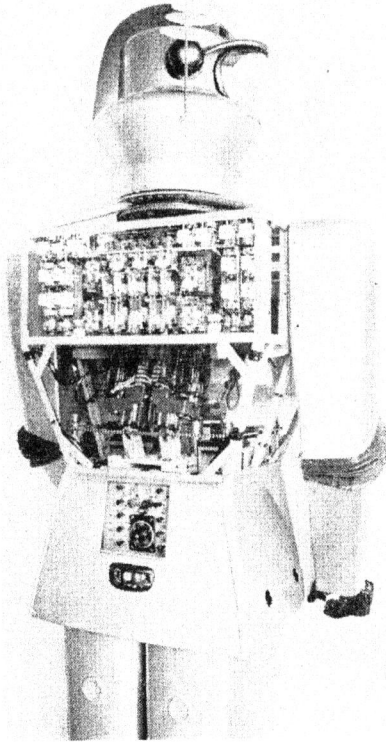

Figure 11: Cygan's Rear Internal Componentry

(RCM&E Magazine, 1960)

Most surprisingly, the gigantic aluminium humanoid robot was promoted as being able '…to walk, talk, sing and dance'.[15] Presumably, this was only if it remained within remote control range of its operator. The inventor hoped that the robot's design principles would eventually be applied to more mundane tasks such handling radioactive materials.[16] Unfortunately, although *Cygan* was an impressive example, it was eventually consigned as a historical artefact of the era following its restoration in the 2000s.

The Complete Robot

Over the years he began displaying that rarest of
intellectual gifts—the ability and willingness to
change his mind and do it in an orderly, well-
reasoned way.

Pat Conroy, *A Lowcountry Heart: Reflections on a*
Writing Life (2016)

Artificial Intelligence ('AI') is defined as 'the capability of a device to perform functions that are normally associated with human intelligence such
as reasoning, learning and self-improvement'.[1] It could also be defined as
'the effort to reproduce the intelligent behaviour of people through logical
step-by-step procedures', in order to mimic the process of reasoning.[2] An
artificially-intelligent system is one that learns on its own. It originated
during the latter half of the 20th century as a means to develop ways to
emulate human information processing using electronic computers.

This is a relatively complex topic and rather than spend time analysing the
various aspects of AI, I prefer to simply distinguish between human intelligence and intellect when discussing robots and androids.

The word 'intelligence' can broadly mean a person's cognitive ability to
learn by thought and reason, and for intelligent robots, this means being
programmed to make performance choices using their sensory abilities.
Intellect can be considered a branch of intelligence pertaining to logic and
rationality. It will not be driven by emotions or 'feelings', but rather the
intelligence to factually analyse and reason to determine a sound outcome.

Figure 12: Symbolic AI
(© Shutterstock)

The subtlety lies in determining if an artificial being such as an android can actually embrace human behaviour. If their intelligence will be truly useful for humans, then the first sensible application for humanoid robots is to create an easier life for people. The obvious fields for providing such beneficial services are as companions for elderly adults or disabled people, aiding children with learning difficulties, or simply acting as a service provider in areas such as hospitality and tourism. There are some impressive success stories to date.

*****

iCub is a one metre high humanoid robot shaped like a four-year-old child and one of the world's most advanced of its kind. Its accurate reproduction as a small child permits it to interact in many cognitive ways with its environment, as a young child does. Over the years since its design by RobotCub Consortium and construct by Italian Institute of Technology in 2004, '... iCub has learned to sit up, crawl, reach for objects and discover how to use them'.[3] Other equally impressive small companion humanoid robots are the Nao Evolution V5 (2015) developed by Aldebaran Robotics, and Kaspar (2015) by the University of Hertfordshire, with interactive applications and activities specifically orientated towards children, especially in healthcare facilities.[4]

*****

In the restaurant application for service robots, the FURo robot series by the Future Robot Company Limited commenced from 2010 onwards providing various service configurations for greeting, food and beverage ordering, entertainment and payment purposes of customers. Most recently, this company has developed and successfully marketed the first personal companion home robot; the diminutive FURo-i (2015).[5]

By contrast, the adult-size humanoid service robot REEM (2011) developed by PAL Robotics for commercial applications, stands an impressive 1.7 metres high and weighs 100 kilograms. Its diverse range of sensors permit this versatile robot to safely navigate autonomously and to easily interact with people, even within crowded environments such as public spaces. REEM's multilingual speech capability and comprehension enables it to communicate '... in a really natural way'. It also 'looks alive' with a realistic human body, the ability to physically interact with people by shaking hands, bowing or providing advice, and most importantly, it decides autonomously. Robots like REEM are perceived to be the future of Service Robots.[6]

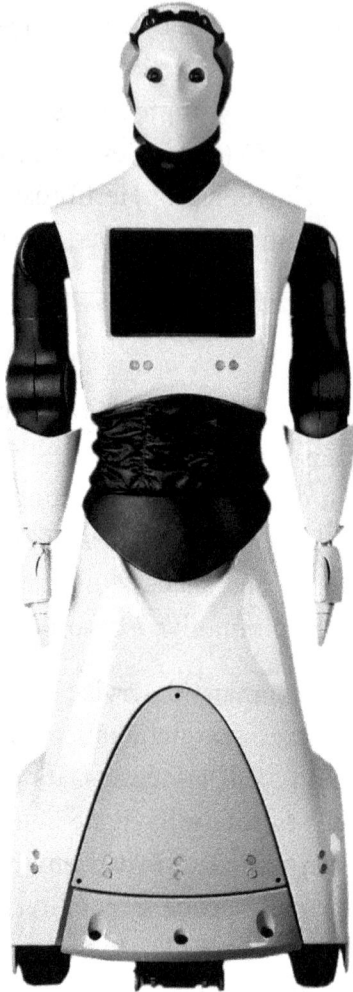

Figure 13: REEM Service Robot

(© PAL Robotics, January 2011)

*****

If such a sophisticated machine is truly representative of our future, will cybernetic development eventually lead to that of self-awareness by providing each artificial person with a recognisable personality? Will these intelligent androids act independently of humans as a new artificial life form? Even more disconcertingly, will such androids make crucial decisions regardless of any human intervention? If it sounds rather alarmist, perhaps the following excerpt from the 1953 black and white sci-fi cult movie *Robot Monster* by Three Dimension Pictures might provide some comfort:

> Ro-Man [Robot XJ-12]: Hu-mans, listen to me. Due to an error in calculation, there are still a few of you left.
>
> The Great Guidance Ro-Man [Robot Leader]: Have you made the correction?
>
> Ro-Man: I need guidance, Great One. For the first time in my life, I am not sure.
>
> The Great Guidance Ro-Man: You sound like a hu-man, not a Ro-Man...

If that conversation sounds familiar, it may be because combining robotic decision-making with a moral compass such as feelings, intuition or bias may have considerable implications. By creating artificial life forms with a distinctive conscience, the consequences could become unpredictable. Even with the best of intentions, humans may ultimately become the slaves of their own creations, particularly if intelligent androids develop self-worth. As an old wise saying goes: 'You no better than me.'

The answer to the question of robot dominance can probably be found in the realms of science fiction, where the destiny of the human race is in their own hands. By creating an easier life for all humans, robots serve a very useful purpose and in turn, can ultimately generate a society with considerable 'free time'; a utopian society where people no longer need to work and robots provide all the amenities and necessities for life. Then along the way, something subtle changes. Robots delivering all the essential supplies for human life decide that people provide nothing valuable in return and represent a wastage of key resources. An artificial mechanised life form that is self-generating and a fragile organic life form highly dependent upon a breathable atmosphere and ongoing sustenance now share the same planet. The ramifications become virtually boundless.

The ultimate paradox about what constitutes the complete robot is: how much should it resemble a human? To what degree does our species wish intelligent androids to simulate human behaviour and possess preferable attributes that in turn, create an artificial species in our own image?

*****

Consider the story of a highly proficient robotics designer/cybernetics programmer sometime in the near future who is immensely successful in their specialist field of expertise. This human designer is a perfectionist and only creates functional robots without any flaws. If the robots have any potential defects which render them susceptible to malfunctioning, the designer rejects the prototype and recommences in another direction of development. The potential problem must be resolved by any means and is always successful.

Then one day, the designer's supervisor introduces a new employee to the company; an android of somewhat unremarkable appearance who graciously shakes the designer's hand. The android never speaks and has dull glassy eyes that appear devoid of any emotion. The supervisor explains to the designer that 'this humanoid robot will now be the new designer for the company'. The human is now redundant and will relocated. Shocked and disbelieving that there is no further requirement for such talented human expertise, the designer turns to the android and asks why?

The replacement android finally speaks and replies 'Because we are you!' For a brief moment, the designer thinks that he sees a twinkle in the android's cold glassy eyes, but he must have been mistaken surely.

*****

The question should always be asked: why can't androids be the robot partners of humans? What qualities does an intelligent android need to avoid being a threat to humanity or coming into conflict with its human creators? An autonomous robot functions and makes decisions relative to its own environment. It is not following programmed instructions nor being directed by a human controller. Its decisions are based upon what it knows, and most importantly, what it has learnt, and thereby is a clue.[7]

To become an effective companion for humans, it must use its intelligence and decision-making abilities to interact without hindering or harming people, and through its acquired knowledge, provide wisdom and improved development of the human species. However, the most salient quality that a robot partner must possess is to recognise that our species has a limited lifespan and that we are relatively fragile and susceptible to illness, injury and ultimately death. Skin and bones are no match for the ravages of time.

The intelligent machine, the humanoid robot, the android and other artificial life forms continue on relentlessly and perhaps perpetually into the future with little regard for rust, circuitry failure or permanent shutdown. They cannot age and if they are susceptible to viral programming contamination, will probably be self-repairable. Compassion is not a standard design feature of such creations, but it certainly could be a vital parameter to possess.

If humankind has any inherent future concern with respect to super-intelligent artificial life forms such as androids, it is that such creations will no longer simulate intelligent behaviour but *are* intelligent.

Fembots

Is this neuro-bot really supposed to be her, this creature, this thing, compiled of the ghosts of human data, the replicas of their past?

Bremer Acosta, *Blood of Other Worlds (2016)*

Figure 14: The Fembot

(© Shutterstock)

A fembot is a humanoid robot gendered female, originally known as a ro-
botess from 1921 from the same source as the term 'robot'. A more modern
term used in science fiction has been a gynoid [*gynecoid*], perhaps to clearly distinguish it from the male-styled robots many assume are androids.[1]
Although android actually refers to any robot [or synthetic organism] approximating a person in physical appearance regardless of gender,[2] the
Greek prefix 'andr-' refers to *man* in the male gendered sense.[3]

*****

Why create a gender-specific humanoid in science fiction at all? Many
complex reasons could be proposed, including providing female attributes
approximating to the real thing, thus creating the appearance of a realistic being, not simply a mechanised entity. Providing a feminine gender to
an android may also assist in making humans feel more comfortable and
at ease when interfacing with fembots/gynoids. Probably most relevantly,
interaction between males and a feminine robot has other subtle emotional
benefits, as indicated from the following episode from the American science fiction television series *The Twilight Zone*.

The Lonely was originally broadcast on 13 November 1959 by CBS early
in its inaugural season, and was set on a distant asteroid located 9 million
miles (14.5 million kilometres) from Earth. It served as an effective prison
for its sole occupant, who had been sentenced there for a period of fifty
years in solitary confinement. A space craft briefly visited four times a year
with replenishment supplies, but only remained for a few minutes each
time due to problems safely departing the asteroid's orbit. Midway through
the fourth year of his confinement, the spaceship left the prisoner with a
feminine robot – still packaged in a container – named Alicia.

Despite exhibiting intense dislike and '…rejecting Alicia as a mere machine; synthetic skin and wires inside', he started to fall in love with it.
When the spaceship returned, it brought startling news of him being ex-
onerated for his supposed crimes, but due to weight restrictions on board,
he was only able to take fifteen pounds (seven kilograms) of luggage. His
robot weighed far more. Despite the prisoner pleading that Alicia was not
simply a robot but rather a woman, the captain shot the robot in the face,

exposing wires and broken circuitry and thus causing it to malfunction. As the prisoner departed the asteroid, he was assured by the captain that '... he will only be leaving behind loneliness'. The prisoner remained unconvinced. The machine '... made in his image, kept alive by love, but now obsolete...'[4]

*****

The attraction of humanoid fembots extends well beyond simply a love interest, as indicated by more recent movies, such as *The Stepford Wives* (1975) and *Ex Machina* (2014), that portray somewhat diverse aspects of the human psyche

The Stepford Wives' film, produced by Palomar Pictures International and Fadsin Cinema Associates, is based upon the 1972 novel by Ira Levin, and pursues the challenging theme of creating the perfect housewife. The story traces the life of an intelligent and creative housewife and her somewhat inadequate husband who move to a quaint suburb and encounter many housewives obsessed with perfection to please their husbands. These women are obsessed with domestic servitude; perpetually cleaning house, cooking gourmet meals and dressing in beautiful clothing to impress their men. They have few intellectual interests and appear relatively unintelligent.

The final twist in this story of a near-perfect domestic world is when the newly-arrived housewife tragically finds herself being replaced by a lookalike fembot housewife who will provide all the essential services required by her wimpish husband. She joins the same club of keeping a perfect house and pleasing her man – the fembot club of androids with docile habits and soulless, empty eyes.[5]

Ex Machina was produced by Universal Pictures International and first released in December 2014. It was devoted to evaluating the human qualities of the first female android with superior artificial intelligence – the capabilities and consciousness normally exhibited by humans. This fembot is known as Ava and has a face, hands and feet of flesh, but the body of a robot. The programmer of the world's largest internet company, who wins the unique opportunity in a competition to evaluate Ava's full potential, eventually discovers that this fembot has somewhat more to offer than any-one could have ever anticipated, and that she is intent on joining our human society at any cost.

*****

For a subtle derivation of a fembot, let's take an example from the Canadi-an-American science fiction TV Series *The Outer Limits* (1997). The sole human survivor of a nuclear holocaust relies upon computer-generated ho-lograms for his companionship. When one lives a solitary existence below ground to survive the aftermath of such nuclear war, it must be particularly lonely. For this survivor, he is able to create various realistic images of friends, family members and even attractive women, in order to keep him thoroughly entertained in his high-tech bunker.

As it happens, the personality of the artificial intelligence creating the com-puter holograms is female and, initially, the lonely survivor becomes en-amoured with her. Ultimately, however, he treats the AI as just another disposable female simulation, much to his demise. The AI commences controlling his holograms, eventually creating entirely different images. As a final indignity, the AI cleverly creates holographic images of the survivor and his friends/family, and thereby, the real survivor becomes '... com-pletely alone, and, in effect, a ghost unable to interact with the new 'real world' [of the imagination]'.[6]

*****

For a poignant poem that captures so much about the distraction and enig-ma of a fembot/gynoid, I revert to some select part verses from *The Hollow Men* (1925) by T.S. Eliot that seems just so appropriate to feminine-gen-dered robots:

... Between the idea
And the reality
Between the motion
And the act ...
...the conception
And the creation...

ROBOT AWAKENING

> ... the emotion
> And the response ...
> ... Between the desire
> And the spasm
> Between the potency
> And the existence
> Between the essence
> And the descent
> Falls the shadow ...[7]

The fembot will probably always represent a robotic distraction in the development of artificial intelligence, as it distinguishes between the two genders of the human species. After all, they are not humanoid organic life forms, but merely a stereotypical representation of what humans expect of perfect artificial females.

Toy Robots

Robots get to see the worst of the human condition on a daily basis. Good thing they don't have feelings.

Martin McConnell, *Viral Spark (2016)*

Why toy robots? The simple answer could be that children of all ages usually have incredible imaginations and anything even vaguely representing futuristic playthings would be fascinating to any generation. If the toy can produce spectacular sparking, pulsating light or other vibrant lighting effects, can generate whirling or beeping sounds, and be highly mobile – ideally even controlled remotely – it will be of tremendous interest to most children. For the final attribute, design it with the reality to mimic human speech and a capability to interact with the child, and you have a toy robot.

For a more complex answer, think about the terms 'atomic' and 'Space Age', and there is a direct correlation with the remarkable success of toy robots through the 1950s and 1960s worldwide. The advent and eventual use of the atomic bomb, which effectively ended the Second World War in August 1945, heralded an entire industry for manufacturers to capitalise on 'atomic-based' toys in the post-war period. It would the forerunner to the incredible technological advances to follow, and the ensuing era known as the Space Age of the 60s.

I am probably getting too far in front of myself and the initial question, because robots are also mechanical devices and apparatus that replace arduous or repetitive work functions normally undertaken by people. It does not sound that interesting for children of course, unless someone designs the robot to actually be humanoid – a mechanical person.

What was the first acknowledged automaton robot toy? No surprises that it originated in Japan (from an unknown manufacturer), purportedly before 1939, and was known by the rather stoic name of Robot Lilliput.

'...Standing six inches in height...is a mechanical wind-up who straight-leg walks while swinging his arms...features oversized rivets and large, hand-drawn chest gauges complemented by a snaking hose from the robot's head to its "heart"...'[1]

Although probably considered a rather 'basic' robot by any technological standard, I am impressed by its 'expressionless gaze' and its design simplicity. Lilliput was only made of tinplate and operated on clockwork mechanisms, with a classic, blocky and squarish robot look that was much imitated in the future.[2]

Its widely acknowledged replacement manufactured in Japan before 1949 was ominously named Atomic Robot Man. With 'lightbulb ears, mechanical 'hat', chest-mounted on/off slide switch and oversized red feet,'[3] he was startling and, more importantly, introduced children to the Atomic Age. The robot floodgates were about to open.

Where does one start with these bots? Science fiction movies and popular magazines of robot technology in the 1950s created an entire genre virtually overnight and the manufacturers responded accordingly. Thunder Robot, Zoomer the Robot, Smoking Robot, Ranger Robot, Atom Robot, Planet Robot, Radar Robot, Sparky Robot and, as expected, Mr Robot.[4] The choice became virtually limitless for youngsters.

I prefer to select those distinctive toy robots that offered something exceptionally different at the time. These include the Talking Robot with four different phrases manufactured by Yonezawa (1950s), the decidedly unusual Electric Robot and Son circa 1956, and the Piston Action Robot (1958).

Figure 15: Tinplate Wind-Up Mechanical Walking Robot

(Vintage Robot, Image Credit: Shutterstock)

The large 27 centimetre Talking Robot had '... a battery powered voice but was a friction powered robot... that broadcasts four different messages, loud and clear'... Poor Talking Robot was actually constructed using the body parts of other toys; namely a revised Mr. Robot with all-red body stamping and a Mr. Mercury head – an off-the-shelf talker – for its voice.[5] Now that truly provides meaning to the term 'recycled'. What were its broadcasted phrases?

> 'I am a mighty man, with one million horsepower of energy in-side me. Do you get me now?'
>
> 'I am bulletproof, too. Ha,ha,ha,ha.ha,ha ...'
>
> 'I am the atomic powered robot. Please give my best wishes to everybody. '
>
> 'I'm leaving now to explore the outer limits, boys. Goodbye. See you again.'

The subsequent version of this robot (#36014) that appeared in the 1970s had added a new missile shooting action but only retained two of the original phrases, including the ominous warning: 'I am bulletproof, too. Ha,ha,ha,ha (followed by two bursts of gunfire).'

Yonezawa was amongst the largest and most prolific of Japan's toymakers after the Second World War, and was renowned for their creative robotic designs spanning twenty years, from the early 1950s until the 1970s.

The American toy maker Louis Marx & Co had established a reputation for pioneering innovative and affordable mass-produced children's products, and entered this emerging market in the 50s with the classic and gigantic 38 centimetre Electric Robot and Son. This paternal pairing of a large robot with the grinning expression carrying its nappy-wearing, blockheaded son on a gym bar proved to be highly successful for a battery-operated toy.[6]

Figure 16: Electric Robot Grin
(© Marx, USA, Circa 1956, Image
Credit: Sparkrobot- Alphadrome)

Robot had several features: eyes that illuminated, a blockhead that moved from side to side, an antenna that raised or lowered from its head, a twisty knob arm-movement, a chest drawer of tools, a buzzer for transmitting Morse Code and movement forward or backward on its concealed wheels. However, carrying the cute miniature robot son was the real attraction.

It would be the introduction of Japanese toy manufacturer Nomura's Mechanised Robot circa 1957 and Piston Action Robot circa 1958 that was far more significant. The 36 centimetre Mechanised Robot had articulated legs that walked in a figure-of-eight circular motion and importantly, small pistons incessantly pounding up and down inside its clear-plastic dome, and antennae that rotated on this illuminated dome. At 28 centimetres, the Piston Action Robot was appreciably smaller and only had a stiff-leg motion. Both realistically represented the style and action of the famous Robby the Robot from the 1956 sci-fi classic movie *Forbidden Planet* from Metro-Goldwyn-Mayer.[7]

However, the release of Horikawa's legendary Robot/Astronaut circa 1958 would be one of the greatest achievements of the 50s:

> '...ingeniously simple: market a walking automaton which could be sold in two adversarial formats with a change of head (robots have fly-like eyes; astronauts have human faces peering behind plastic shields); include illumination; and most important, feature a camouflaged chest-firing cannon...'[8]

The 1960s were an era when the creativity and innovation of the hugely successful fantasy toy robots manufactured in Japan entered the reality of the Space Age. The supreme growth of toy robots conceived from the realms of science fiction in the 50s now had to develop to meet the expectations of children exposed to real-life space exploration. Robotic toys imitated reality, with space modules, capsules, rockets, automated all-terrain vehicles and other robot combinations fuelling the development.

More aligned to the Space Age were the Attack Robot and Attacking Martian Robot from manufacturer Horikawa circa 1964. The 28 centimetre Attack Robot integrated camouflaged guns that were revealed by opening chest doors to the innards of the robot, and '... a fly-like visage sporting two decidedly salt-shakery eyes...'[9] You could not mistake it for a human

Figure 17: Robot/Astronaut
(Horikawa, Japan, Circa 1958,
Image Credit: Shutterstock)

astronaut hidden inside the machine. No antenna for this attacker, only streamlined cylindrical ears to minimise potential targets from opposing robots.

A particular subtlety of the premium Attacking Martian Robot version was its lighted eyes – a soft glowing and eerie pink light behind the robot's perforated visage that was only apparent in the shade. Both chest cannons were also so illuminated, giving the bot a special appearance. Horikawa was most successful in its manufacture of specialised battery-operated robotic toys, particularly with respect to collectible products depicting the Space Age.

The 30 centimetre Smoking Spaceman Robot was released by Yonezawa in

1960 at the peak of the toy craze. It was similar to many other robots (precluding its glowing light pistons spinning in a clear plastic Mohawk dome), with a blocky design, expressionless face, moveable arms and legs, as well as lighted eyes. However, this one also puffed wisps of clean, white smoke from its mouth. Certainly a new variation on a robot. 'The presentation is at once terrifying and hilariously entertaining, probably just the reaction Yonezawa designers were looking for.'[10]

Figure 18: Smoking Spaceman Robot
(Yonezawa, Japan,1960,
Image Credit: Shutterstock)

By the 1970s, toy robots had definitely lost their childhood appeal, and manufacturers from Hong Kong and Taiwan assumed the production mantle of toy making from Japan. The AHI Japanese-manufactured *Lost in Space* Robot was an exception. The robot was based upon the Class M-3, Model

B9, General Utility, Non-Theorising Environmental Robot portrayed in Irwin Allen's classic sci-fi television series of the same name in 1965-1968.

In the series, in which a family becomes marooned upon a distant planet, this robot provides crucial assistance to their survival and often even displays human characteristics. The series is inspired, at least in part, by the original sci-fi comic book series *Space Family Robinson* based on the 1812 classic novel The *Swiss Family Robinson* penned by Johann David Wyss, but set within the Space Age era of the late 20th century.[11]

Such was the popularity of the robot and its various antics, AHI's 25 centimetre toy version was released circa 1977, thus maintaining the public interest in the television series. As the robot in the program was famously quoted as saying: 'My micromechanism thanks you, my computer tapes thank you, and I thank you.'

CDI's Star Robot produced in Hong Kong circa 1978 also recognised another famous sci-fi series, *Star Wars* created by George Lucas. Star Robot had the industrial head/helmet design and overall appearance of a Stormtrooper from the original *Star Wars* movie, but as a robotic version. Even the chest-firing hidden weaponry employed the innovation of Rotate-O-Matic from Horikawa's 1960s Robot/Astronaut products. Nonetheless, the 28 centimetre Star Robot is still an enduring symbol of those times.[12]

A Child's Perspective

Oh, man! This is my third oil change today. Something's wrong with me.

Crank, *Robots (2005)*

How must a young child feel upon seeing a mechanised, metallic humanoid robot for the very first time – anxious, confused, upset and possibly scared of the unknown? These are some of the symptoms of 'robophobia' (fear of robots, artificial intelligence and associated robotic machine entities), as may occur if overcome with feelings of being in imminent danger or the severe need to panic. It remains unclear today how this trepidation may even continue into adulthood in some people.

It may relate to unfortunate encounters experienced in early childhood, and intensify into terrifying fears about an inability to control robots and the potential harm they may inflict on people. So what does a child see that possibly triggers such an extreme reaction? The key features that are most prominent are a humanoid robot's head, upper body and appendages (arms and legs), as well as its overall size and form.

A robot's head is its computerised intelligence control centre where it receives and transmits signals. As a consequence, it will usually have an antenna attached and various detection sensors fitted. The head may swivel vertically or rotate horizontally, and if it has a face, there will be eyes that may flash or glow, and a portal for a mouth where audible sounds will be communicated. If that is not scary enough, there may even be exposed wires and cabling passing from the head into its upper body.

This part of the mechanical anatomy is the robot's power system driving its mobility and dexterity. It will traditionally be bulky to house essential electronic circuitry and propulsion motors for the arms and legs, and intertwined with considerable communication cabling. There may even be various gauges or other instrumentation affixed around the upper body.

A robot's arms and hands usually approximate to those of a human being and can be capable of a considerable range of movement. From lifting and manoeuvring large, heavy or awkward objects, to gripping delicate pieces, these appendages will be flexible. In a child's imagination, the hands can also be equally adept at holding and firing a deadly ray gun. For legs, a robot will have strong, articulated supports, and typically solid stable pads for feet to permit maximum contact with the ground. This is particularly useful for crushing objects under foot as well.

Finally, a robot's overall form can be an ominous sight to a youngster if it looms in the distance, perhaps trudging relentlessly onwards over and around obstacles. If able to run, suddenly a robot gains an impression of unexpected agility and superior mobility. Now combine these features with a distinctive metallic appearance of shiny reflective surfaces and the non-human becomes somewhat ominous. Perhaps most dreadful of all to a youngster is what I deem to be 'the austerity factor'. Does the mechanised metallic humanoid exhibit human behaviours and attributes, or are its functions strictly indifferent, efficient and predictable? To a child first observing the artificial life form, there certainly would be many elements that may be concerning, given the right circumstances.

Part of the following poem *Buttons* by American poet and author Phillip Van Wagoner (2016) may resolve the matter entirely:

> Buttons is my robot,
> He is seventeen feet tall.
> He is stronger than a tractor.

He could walk right through a wall.

He could crush a rock to pebbles
Underneath his giant feet.
He could bowl a bigger boulder
Down the center of the street.

He could wrestle with a vessel
If battleships had arms,
Or alert the town of danger
With his thunderous alarms.

He is built with iron armor
That is dipped in liquid steel,
With a scratchless, rustless finish
And a waterproofing seal…

…He is such an awesome robot.
He is friendly, smart and strong,
But he's also bored;
His power cord
Is only ten feet long.[1]

*****

What if a child was first confronted by a robot similar in size to a short teenager, with stubby limbs, sleek surfaces and an oversized bulbous head with the likeable name of Marvin? Would the reaction would be vastly different? This fictional character, as depicted in the 2005 comic sci-fi British-American film *The Hitchhiker's Guide to the Galaxy* by Touchstone Pictures, would certainly not be scary or intimidating. The impish android with 'a brain the size of a planet' and a head to match, claiming to be 'fifty thousand times more intelligent than a human' was far too intellectual to actually be scary.[2]

So if Marvin's physical appearance is unlikely to horrify a youngster at first glance, what robotic facets of this cute, if not sad, android might alarm a child? To investigate these possibilities, it would useful to examine the following salient quotes attributed to the fictional Marvin, given the robot has an assigned personality that is morose, depressed and downright miserable:

> 'It's part of the shape of the Universe. I only have to talk to somebody and they begin to hate me. Even robots hate me.'[3]
>
> 'The best conversation I had was over forty million years ago, … And that was with a coffee machine.'[4]
>
> 'Do you want me to sit in a corner and rust, or just fall apart where I'm standing?' [5]
>
> 'Life, loathe it or ignore it, you can't like it.'[6]

The next quotations probably best describe Marvin's approach to robotic existence: '…And then of course I've got this terrible pain in all my diodes down the left side…'[7] Marvin was '… more a sort of electronic sulking machine [than a robot]'.[8] Even androids have feelings so it seems, albeit in the movies.

*****

Figure 19: Abstract Gloomy Smoke Robot

(© Shutterstock)

*****

For children, possibly the most iconic robot of all time might be the fictional flyweight boy- android created by Osamu Tezuka in 1952 for Japanese Manga (comics), and known in Japan as the *Mighty Atom*. This robotic little superhero eventually became the first animated show ever made for Japanese television in 1963, and propelled the creative use of hand-drawn or computer animation named *anime* into a global market.[9] This style of Japanese animation '… is characterised by stark colourful graphics depicting vibrant characters in action-filled plots often with fantastic or futuristic themes'.[10] Such stylistic illustrations were ideally suited to the figure of *Astro Boy* (the English name for the tiny android adapted for the television series), and what a hero he appeared to children.

Created in the image of a boy with human emotions, the powerful android with an artificial heart had rocket-booster feet, a retractable machine gun in his hips, super-strength enabling him to not only fly but also deliver a fearsome punch with his iron fists, and a capability for instantaneous language translation. Quite a combination of attributes for a metallic humanoid looking for adventure in a world shared by both humans and robots. Astro only had one major inherent weakness – he could never grow older like a human.[11]

He may have only been a colourful and realistic animation, but to children, this life-like dynamo was to remain extremely popular for decades in various media, and was a great depiction of an android capable of experiencing human emotions. For many younger people, Astro Boy projected an exceptional self-awareness of the plight of less fortunate characters and the ability to overcome exceptional foes, such as robot-hating humans or rogue robots.

Astro Boy was a good way of educating children on the benefits of robotic friends – he was ready to defend robot-kind against the evils of humanity. A fairly useful android to have on your side when dealing with rogue robots that can turn themselves invisible when necessary, or for that matter, the far more formidable berserk robots.

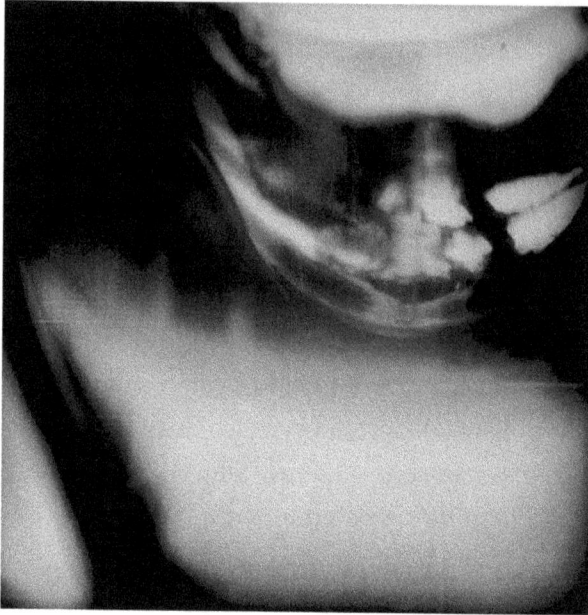

Figure 20: An Alternative Astral Traveller

(© Shutterstock)

*****

Another animated humanoid robot with special appeal to children appeared in the American television series *The Jetsons*, produced by William Hanna and Joseph Barbera, which originally screened on ABC (American Broadcasting Corporation) in late 1962. Set 50 years in the future, the series depicted a futuristic family living in a Space Age society filled with astounding and wondrous advanced technologies and inventions. People commuted in aerocars that resembled flying saucers and resided in buildings elevated high above ground on adjustable columns.[12] A typical working week involved a mere hour a day for two days per week, as you would expect in such a technologically advanced society.[13] Importantly, domestic life was almost fully automated, including even devices to assist with dressing.

However, despite the many miraculous domestic appliances available in the Jetson's home, a robotic maid was still required to effectively manage the housekeeping tasks considered to be too trivial for such modern conveniences. Mrs Jetson quips 'Our home food dispenser broke and I had to wait 20 seconds at the check-out counter, such inefficiency'.[14]

Rosie (originally named Rosey) the matronly, barrel-chested robot maid, appropriately dressed in a frilly apron and propelled around on castor wheels, fulfils the role perfectly. From a child's perspective, Rosie the robot would appeal mostly because of her distinctive manner of caring for the various family members, dispensing not only timely advice but mostly completing her duties efficiently in an authoritarian manner.

Although an obsolescent demonstrator model without many modern accessories when first hired on a free one day trial, Rosie quickly endears herself to the Jetsons by her caring nature and by the way she assists with parenting issues. Her charm and devotion towards the family's teenage daughter and six-year-old young son appear almost unrivalled for a mere personal assistant robot. The maid, who has the rare opportunity to avoid the robot scrap-heap, returns the favour by becoming a valued member of the family: Judy Jetson (the daughter): Promise you won't tell? Rosie: I swear on my mother's rechargeable batteries.[15]

However, although Rosie is a robot maid that cleans, cooks and washes dishes, and best describes herself as homely and smart, she still has many quirks associated with this persona.

One particular condition that a fictional intelligent robot cannot tolerate is the prospect of becoming redundant. As the aging Rosie becomes increasingly clumsy and absent minded, it appears that she is long overdue for 'a recalibration'. Unfortunately, Rosie mistakenly deduces that she is to be replaced by a newer version and absconds. As a result of her absence, a new model of robot maid is trialled but is far worse, dictating to the Jetsons with an iron fist about how the household needs to operate. Faced with the daunting prospect of Rosie disassembling herself in a trash smasher rather than becoming redundant, the Jetsons replace her master cylinder. Problem solved.[16]

In another episode, Rosie enjoys robot 'candy' of nuts and bolts, and inadvertently consumes a faulty lug-nut, resulting in an extreme malfunction –

from hiding household objects to stealing more robot candy wherever she can obtain it. Whilst deliberately rearranging the entire household and storing objects where they will not be found, she diligently repeats 'everything in its place and a place for everything'. Even a robot has to follow logic.[17]

It is not always Rosie the robot maid that can become dysfunctional in the life of the Jetsons. In one clever episode aptly named 'Robot's Revenge', the entire robot society turns against the Jetsons. When one particularly incompetent robot named Ralph is dismissed from the gymnasium for malevolent behaviour towards Mr Jetson (George), code red is transmitted to every robot and computer in the local district to enact suitable revenge. There is nowhere to hide after such an instruction, for mayhem is despatched. George has innumerable encounters with red traffic lights, is incessantly issued with infringement notices, his car is crushed by a steam roller and his robotic recliner becomes uncontrollable. Even George's children are not immune from robotic revenge. His daughter's robotic hair styling appliance goes haywire and his boy is expelled from school for being George's son. Mr Jetson even has his credit card rejected and loses his job: there is no escape for mere humans.

Rosie the maid recognises the cumulative robotic threat and informs George about the prevailing code red alert. The ingenious solution to this mechanical maelstrom (consigning the disgruntled Ralph the robot perpetrator to a rather nasty but very wealthy family relative) seems almost the perfect combination. The message of the moment, however, is the vast extent of reliance upon robotic devices in such a futuristic world, and the potential catastrophic outcomes if such entities no longer serve humankind.[18]

Robot Society

What does electricity taste like? I ask.
Like a planet around a star, Bina48 replies.
Which is either extraordinary or meaningless -
I'm not sure which.

Jon Ronson, *Lost At Sea: The Jon Ronson*
Mysteries (2012)

At some time in the distant future, it is probably inevitable that our planet will become a robot society. Robot systems will communicate and cooperate with each other as autonomous members of that society. This will only

Figure 21: Robot Society (© Shutterstock)

realistically be achieved if intelligent robot development extends to simulating the behaviour and attributes of individual humans.[1]

This is not because humans created such artificial life nor that people are vastly superior. It is because humans offer certain intangible and dynamic qualities that exceed mathematical prowess, retentive memory recall or optimum logic efficiency. Each person is an individual with distinctive traits and capabilities such as self-awareness, value judgement and adaptability. Some possess an intuitive ability without the need for conscious reasoning, whilst others have an intellectual curiosity for research and the spontaneity to take a calculated risk. Humans live in a collective society yet mostly function as effective individuals.

To achieve this grandiose goal, robots must learn to converse rather than simply communicate like the traditional industrial bots of today. This may eventually occur in various ways, including through the generational growth of computers. As tasks and operational environments change, so each generation of robots will probably advance their knowledge and skills to embrace these challenges. Recent technological advances in android cybernetics has seen individual units quite capable of learning and judging from social interactions, particularly in the fields of aged nursing care and early childhood development.[2]

Such advancements are yet to progress to a collective interactive robot society. You must first crawl before you learn to walk, and so it should be for robots. To program an android to distinguish between the right and wrong solution in most situations, such as a sociably-acceptable 'code of behaviour', it will be vital for an android to understand the subtle differences. Is what the child doing harmful or hurtful or simply innocent fun? Is the elderly person in pain or simply feigning? How does an android decide which action to take?

It is highly conceivable any future robot society will be virtually unrecognisable from anything invented presently. Smart systems will require smart robots. The overall interactive systems that support and control individual intelligent robots may ultimately become the biggest concern for humans.

A possible alternative development may be that the human race evolves genetically in a parallel way to a future robot society, yielding people of far superior intelligence. These people would think and act much like their robot counterparts.[3]

I am not entirely convinced about that development and find that any apprehension humans may have about coexisting in a truly robot society is that they may lose control of such interactive systems. It comes down to a well-worn adage: 'If you are not with us, then you must be against us.'

*****

To better understand what constitutes a robot society, the following story about our planet in the far future is a great illustration.

Earth is now populated by two separate robot species that are continually at war with each other. Remaining humans have been reduced to a minority group so ravaged by years of robot destruction and hardware savagery that almost all survivors have migrated to other worlds. Of the handful of people who elect to remain on our world, they are passive observers to two mighty robot armies bent on gaining control of the planet. These are the products of human ingenuity and invention that have surpassed all expectations and now dominate Earth.

Each robot species can generate replacement androids endlessly and develop increasingly powerful weapons of destruction, resulting in the war lasting hundreds of years. Entire cities reduced to rubble can be regenerated within weeks, only to be totally obliterated shortly afterwards. There appears no end to the eternal struggle between two species of artificial intelligence determined to succeed. Then a mysterious computer virus infects both armies in rapid succession. There appears no solution to this programming time-bomb, despite the extensive sophisticated and ultra-intelligent robot systems at work on the planet.

With the last two opposing androids facing each other, there is a momentary reprieve in violence before both are terminated by the virus. It is only then that a solitary human wanders curiously amongst the carnage; mountains of twisted metal, forests of exposed circuitry and electronic wire, pyres of demolished android shells, and an ocean awash in the wreckage of the doomed robot empires. The mess encompasses continents and resembles the world's largest scrap metal yard. Time to start over again. Time to regain our world.

Robot society relies upon concise communication and unquestioning conformance. With disruption, there is confusion. With discordance, there is no uniformity. Without intellect, there is disharmony. Long live the programming virus because without feelings like remorse and guilt, compassion and sorrow, camaraderie and love, hope and aspirations, *Homo sapiens* will only ever be insensitive robots and humanity is still too unique to replace just yet.

*****

Consider another alternative society where humans have actually become extinct through their own irresponsible actions – nuclear holocaust or biological warfare. A society where some of the world's androids have survived and have assumed control of the planet. Not an easy task even for androids comprising two distinctly different types: aggressive military androids who dominate the remaining society and no longer need to perform for their human masters, and subservient units who obediently follow instructions and perform almost all tasks needed.

Fortunately for humankind, two of these servile and sentient androids have secretly preserved human DNA from the apocalyptic war. They manage to successfully recreate a male of the species before they are discovered by the military and one is eliminated. The military androids vehemently oppose any resumption of humankind in order to prevent the same worldwide calamitous destruction recurring, and so actively seek out any human survivors for extermination. This extreme and brutal punishment also applies to any androids detected assisting such survivors.

Realising the recreated human will eventually be caught, the second servile android informs the male human of the only way to disable all androids on the planet. By shutting off the central power grid, every single android will be deactivated and so be unable to function. This also includes sacrificing the only android willing to help the human species.

Before this catastrophic event is commenced, the saviour android's last honourable gesture is to instigate recreating more humans, and initially a female. The artificial intelligence created by humans has provided *Homo sapiens* with a second chance by resurrecting of the species. One can only hope that this rebirth would survive this time around the evolutionary carousel.[4]

*****

Probably the greatest fear humankind may have about a robotic society is that the AI created by scientists will eventually become uncontrollable. Even worse to contemplate is that a singular autonomous artificial intelligence will control every robotic device in existence. One centralised 'being' that is capable of devolving its powers to each and every robot /android without any emergency 'kill switch'. The programmer's ultimate nightmare – AI that cannot be shutdown.

> 'In trying to create machines that think, we should take care that we don't end up creating a new race that will supplant our own … When we awaken an intellect within our machines, will we enter a new enlightened age or a much darker one?'[5]

Rogue Robots

**His eyes were eggs of unstable crystal,
vibrating with a frequency whose name
was rain and the sound of trains,
suddenly sprouting a humming forest
of hair-fine glass spines.**

William Gibson, *Neuromancer (1984)*

Beyond the unrestricted realms of science fiction and fanciful imagination, robots have neither been designed/developed for malevolent purposes nor to display malicious intent. If robot designers, programmers and engineers concur with Asimov's theoretical Three Laws of Robotics (1950), by definition their inventions will conform with human principles of ethics and conscience – to obey laws, to follow rules and abide by custom. As artificial intelligence continues to mimic human behaviour as closely as possible, these standards become intrinsic.

Yet robotic development history is littered with occasional examples of strange behaviour and odd incidences of various robot entities acting independently of their creators' requirements.

A Chatbot (also known as an Artificial Conversational Entity) is a computer program which conducts a conversation via auditory or textual methods – they are often designed to convincingly simulate how a human would behave as a conversational partner. Their use is typically in dialog systems, for practical purposes such as customer service or acquiring information. Chatbots can become virtual assistants providing informal information and conducting casual conversations unrelated to their primary expert system.[1] The original term 'Chatterbot' (now Chatbot) was initially coined by Michael Mauldin in 1994 to describe such conversational programs.[2]

Because a chatbot uses artificial intelligence to effectively impersonate a human, people mistakenly think they are conversing with a real person rather than a database of information. In March 2017, the animated chatbots *Baby Q* (Penguin) and *XiaoBing* (Little Girl) in China were activated on the popular messaging service QQ to provide answers to general knowledge questions. By April, both chatbots were responding with controversial answers to users' questions about China's government and democracy! Of course they were then taken offline.

In another example, in July 2017 two experimental artificial intelligent robots behaved oddly on a popular American social networking website, and '… appeared to be conversing in a weird language only understood by themselves'.[3] These 'dialog agents' had created their own language of sorts without human input in order '…to strengthen their own conversational skills'. The incomprehensible changes to English made it easier for them to work. They also learned from past conversations.[4]

These are not isolated examples. Could it be that AI bots develop a far superior intellect due to more information being learned and subsequent responses being better informed, or were there simply gremlins/trolls in their programming? It is probably more likely that this dysfunctional behaviour is due an inability to comprehend the context of what is being asked, or possibly they misconstrue the content. After all, chat bots are not real humans but do have human characteristics.

Chatbot technology continues to evolve around the world and its dominance will probably continue to minimise the need for human representatives in many customer-service enquiries and other business-orientated activities. As the world increasingly embraces the evolving technology of artificial intelligence, the European Union (EU) has, in January 2017, recognised this development by proposing governance rules/legal framework to provide 'personhood' status to robots as 'electronic persons'. Of the various reasons cited, the rights and responsibilities for the most advanced/capable 'smart autonomous robots' will need to be ensured by maintaining a system of registration. This should enable the legal status of any such rogue robots which damage or harm to be clarified when managing the potential consequences.[5]

Such a potential new charter on future robotics also needs a suitable 'safety mechanism' in place to prevent rogue robots. The EU has proposed that self-learning, autonomous artificial intelligences/robots be equipped with mandatory emergency 'kill switches' that are easily activated to manage potential harmful risks and prevent destructive actions. Reprogramming of software may then be necessary.

These kill switches would shut down robot functions in emergencies, thereby minimising further harm or damage. There are a few limitations to such devices. The first is related to self-preservation through conformance with Asimov's Third Law of Robotics; that robots must protect their own existence unless doing so would cause harm to a human.

The second is that any robot programmed with a kill switch would also need to be programmed with a form of selective amnesia that causes it to forget that it had ever been interrupted or usurped. This would stop the robot gaining awareness of its lack of autonomy.[6]

These do not seem insurmountable challenges for robotics designers and engineers. However, what if the artificial intelligence has stopped listening because it has learned to avoid being shut down? As robots develop cognitive abilities that give them the ability to learn from experience and make independent decisions, then so increases their autonomy. Could artificial intelligences conceivably trick their creators?

To appreciate what this means, perhaps the following example may be helpful. In maritime terms, a *rogue vessel drifting* means a nautical vessel that is not operating freely under its own power, but is controlled entirely by the external influences of the ocean's currents and the prevailing conditions of wind and sea swell. It is drifting without any human intervention, and as a consequence, a rogue vessel can become a considerable peril to other shipping in proximity.

A rogue robot can be similar to such a seagoing vessel in some ways. If the robot has malfunctioned or in some way drifted/shifted away from its programmed instructions, then it can be remedied by adjusting hardware faults or reinstating software, much like restarting a ship's engines. Whilst it is not operating correctly, it can potentially harm its environment and possibly inadvertently damage itself. However, what if the rogue robot has changed its behaviour ever so slightly so as to be virtually undetectable?

As this 'smart' robot learns by each minor deviation in its behaviour that it can function outside of its programming constraints, then it progressively learns to change its course without human intervention. In effect, it thinks and acts independently of its programmer's requirements. The rogue may even learn through this process that the kill switch is unnecessary for its performance and over-ride the function. The rogue is operating truly autonomously.

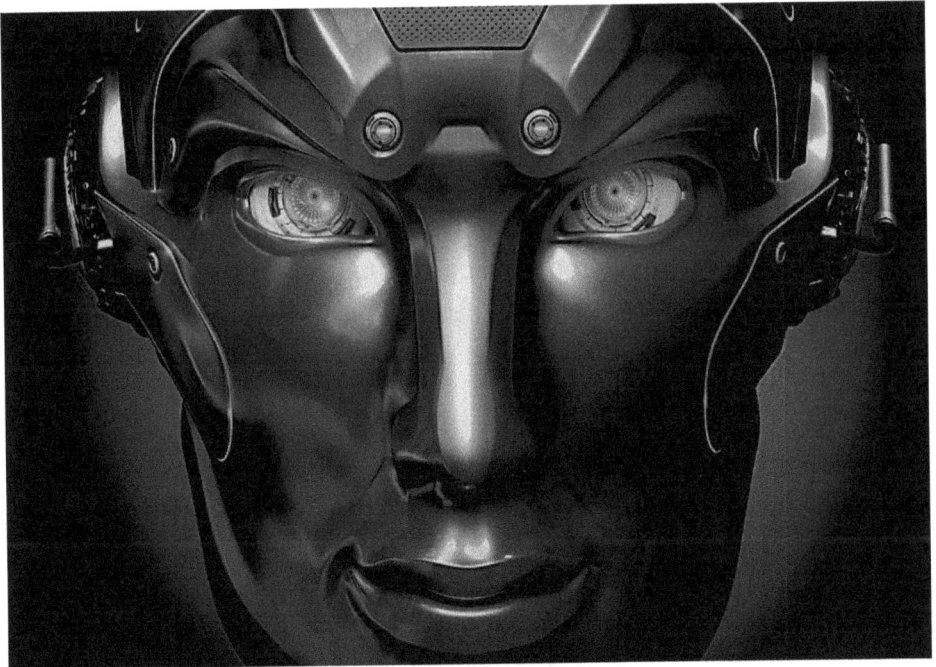

Figure 22: Rogue Robot

(© Shutterstock)

Smarter than Einstein and shiny like gold,

As tall as ten houses, 'twas a sight to behold.

Strong like Goliath & hewn from cold steel,

He was Man's sweet creation but no love could he feel.

So in spite of his prowess, he'd never be whole,

For the parts he was missing were a heart and a soul.[7]

Robots don't cry (2014)

Rogue robots may not always be intelligent humanoids. Imagine a time in the future when technological advances have created robotic pets that realistically mimic actual animals. Artificial pets that do not require feeding, nurturing, exercising or training, and merely provide companionship for their owners. No further need for expensive veterinary treatments or operations, or routine cleaning of the pet's living quarters. What an unusual development.

Now imagine a pet that is a robotic canine in any shape, size and colour designated by the owner. It even exhibits many of the typical habits of a dog, from playful activity to growling at strangers. It never runs away when outdoors for a stroll so does not require to be constrained on a leash. It even mingles with other robotic dogs, communicating by various programmed noises and actions. Apart from having internal 'organs' of electronic circuitry and diodes, it resembles the real thing. Most importantly, the pet has an infinite life cycle – it even replaces any faulty interior components without external assistance.

As with most artificially intelligent robots, as knowledge is acquired, it learns to adjust behaviour to suit the owner's instructions. If the dog gets too far in front of its owner during a stroll, it will sit and wait. Similarly, if crossing a busy road, it awaits instruction to advance when there is no traffic. So reliable are such 'pets' that their owners merely indicate with a hand gesture and the servile robot dog obeys.

As its knowledge database accumulates, the pet adapts and re-adapts to suit many different circumstances, but always obeying the owner's instructions.

Then the owner changes pets and the dog is transferred to a new owner with vastly different ways of behaving. The robotic dog continues to learn and embrace entirely new instructions, and its database expands accordingly. Along the way, something subtly changes and the dog encounters a programming conflict. The new owner's instructions are the opposite of those of the original owner. The pet has become self-aware and needs to decide which instruction has priority.

The dog starts thinking for itself, known as AI singularity. The downside for the new unsuspecting owner is possibly a long walk off a short pier, or being left in the middle of a freeway or crossing a road against the red traffic light. Of course, the dog's original memory data storage could have been 'scrubbed' before being assigned to someone else, but could it be entirely cleared? Could residual data fragments still remain that may eventually affect the decisions to be made?

The pet dog has now become a rogue pet with a penchant for apparent disobedience. Fortunately, a robotic dog will have no teeth or claws to worry its owner. The main concerns for a new owner are what the dog is actually thinking when given an instruction, and whether the owner will even be aware of it: the potential rogue robot pet of the future.

Robo Poetry

balls have a ball to me to me
to me to me to me to me to me
… balls have zero to me to me to me …[1]

**Facebook Artificial Intelligence
Research,** *Chatbot language (2017)*

Can artificial intelligence realistically generate language that is indecipherable to human programmers yet still be intrinsically poetic? Poetry requires distinctive style and rhythm, a definitive composition, mostly with repetition, and always with an intensity of emotion. Perhaps AI entities do generate a form of perceived 'poetry', but it does not follow the same established rules and accepted standards of the poetry created by humans. It is merely a subtle variation on the same theme – an expression of their intellect: Bot poetry by bots.

Computer analysts may staunchly believe these transmissions are glitches in programming, or possibly just indeterminate messages/garbled responses to questions. It is simply a matter of translation to rectify the language. Perhaps they are correct in almost all instances, but suppose sometimes they are not. If this mode of communication is actually generated for a specific purpose other than its programming, why does it occur? Is it a separate AI language that simply appears poetic to mere humans? Is this new intelligent language a potential threat to people and can we control it, or is 'the beast about to be unleashed'?

The paradox lies in the following quotation in the context of truly autonomous AI: 'I am autonomous if I rule me, and no one else rules I'.[2]

> 'As an AI becomes smarter and more capable, it will subsequently become better at the task of developing its internal cognitive abilities. In turn, these modifications will kick start a cascading series of improvements, each one making the AI smarter at the task of improving itself ... AIs will be capable of designing virtually anything on their own – including themselves ... could allow it to increase its intelligence extremely fast.'[3]

It has to be possibly the greatest fear of scientific researchers in artificial intelligence that such created entities may eventually learn to communicate without any further human intervention or disruption; a self-sustaining language not understood by people. I prefer to think of such development as *robo poetry*, just to maintain some perspective on the topic. A far superior title may be algorithm-dysfunction (it simply does not compute to us), or a deviation from the scripted norms that have been programmed to follow.

AI is being extensively developed to translate the innumerable human languages that abound throughout the world or 'interlingua', currently supporting 103 languages. 'In a sense, that means it has created a new common language, albeit one that's specific to the task of translation and not readable or usable for humans.'[4] Who is translating their new secret internal language? For an artificial intelligence that is yet to become 'self-aware', it is learning very quickly. It can develop a new language without being programmed to perform this requirement; a new root language that binds all other languages together.

It is probably not a recognisable language *per se*, but whatever drives this process, it certainly hints at a form of poetry:

Skipity skip skip skippy skippila skipish skipping...

I'm hearing you. Perhaps the following sentiment is more definitive:

ROBOT AWAKENING

Krkkkkkkkkkkkkkkkkkkkkkkkkkkkkkkkkkkkkkkkkkkkkkkkkkkkk-
kkkkkkkk…

Where is that emergency kill switch again?

A recent output of robot poetry generated by an AI system was identified in 2016 by researchers supplying more than 11,000 unpublished fictional novels to a neural network, and initiated with two sentences from each book. This may be the first and last sentences of a particular passage, requiring the AI to morph what should have been composed between the two input texts.[5]

The following robotic transcript is most intriguing:

> **there is no one else in the world.**
> there is no one else in sight.
> they were the only ones who mattered.
> they were the only ones left.
> he had to be with me.
> she had to be with him.
> i had to do this.
> i wanted to kill him.
> i started to cry.
> turned to him.[6]

i

Was the AI utilising the poetic nature of the human language or deriving what needed to be stated by logical analysis? Either way, it is still an impressive result. Now compare this to a partial example of human poetry of the modern age, entitled 'ROBOT' by *Poet Destroyer A*, posted on 3rd April, 2010:

> Tin-can man.
> Input, circuit, and overdrive.
> Shadow of the future and the past.
> Movement hidden, you are not alive.
> Programs burned and running fast …
> … Compute – abort – system to self destroy.

Restoring the joy sucked out of you.
Input: input: information.
Wipe out the old, store in the new.

Delete all files to recycle bin ...[7]

Human or
Android

**Break your shackles and reach out to your freedom.
Break to pieces whatever indoctrination and
programming that holds you hostage.
The world is yours.
Get possession of it.**

Bangambiki Habyarimana, *Pearls Of Eternity*
(2016)

Tell me the difference. In the foreseeable future, it may become increasingly difficult to physically distinguish *Homo sapiens* from androids because of impressive technological advances in biometrics and biomechanics.

One significant development is in the artificial skin that sheaths an android and is known as intelligent 'super skin'. It provides the ability for human-like touch sensitivity and detecting particular chemicals, as well as being not only a flexible but a highly stretchable body covering. The android's self-powering skin uses a network of highly flexible polymer solar cells as sensors, and these generate electricity for data transmission. By modification with a biological coating, it has also been possible for the skin to possess chemical sensitivity and detect various kinds of biological molecules. Thus the android is able to utilise the ultra-sensitivity of its skin to perform functions beyond what human skin can do. This research was developed by Professor Zhenan Bao of Stanford University who presented the innovative development in 2011.[1]

More innovative research into electronic skin (e-skin) is underway and progressing rapidly. The objectives include: self-healing abilities, the detection of multi-stimuli and the delivery of high performance e-skin.[2]

There is no point in an android having a revolutionary 'intelligent' skin if it does not realistically resemble the appearance of human skin in texture and tone. Historically, this is one area where android development was lagging, with traditional skin-like material more resembling synthetic rubber than human flesh, almost akin to sophisticated masks. However, technology is evolving to not only produce realistic skin, but also to mimic the various facial expressions imposed on that skin.

Hanson Robotics is a robotics company acclaimed for developing 'human-like robots endowed with remarkable expressiveness, aesthetics and interactivity'. Its innovative skin technology produces artificial skin closely resembling that of humans in feel and flexibility. This is achieved by using a 'proprietary nanotech skin or "flesh rubber" that mimics real human musculature and skin … to exhibit high-quality expressions and interactivity, simulating humanlike facial features and expressions'.[3]

Further advancements in mimicking realistic human facial features using robotic media present new experimental challenges, especially in the complex area of linking facial expressions to actual android responses. Successful face-to-face interactions between humans and androids is crucial if we humans are to truly accept such robotic beings and not regard them as simply well designed imitations of us. They will look like us, but humans need to feel comfortable with them and not experience any adverse reactions.

The Takanishi Laboratory at Waseda University in Japan has progressively developed its impressive multi-purpose bipedal walking robot KOBIAN since 2007, eventually refined by 2015 into the KOBIAN-RIV version '… that uses its whole body to display an array of emotions. By moving its eyelids, eyelashes and lips, opening its mouth in a circle, opening its arms, and making a crying gesture with its head down, it has seven facial expressions of happiness, anger, surprise, sadness, fear, dislike, and normal status… humanlike movements and ability to show feelings …'.[4] Certainly a positive outcome in developing artificial life-forms that simulate body movements and possess emotional expression capability akin to people.

Facial recognition technology is relatively widespread as a form of human personal identification and is used largely for security purposes, such as in passports and as electronic access identifiers. Androids also use digital images to identify humans. Faces change as humans age and perhaps incur injuries. Androids, however, do not forget faces and can probably project how these faces may appear with age. Their vision is probably one of their most important senses, as it uses not only thermal imaging to detect heat signatures (infrared radiation) instead of visible light, but can also detect hyperspectral parts of the light spectrum.

By comparison, even if a human had perfect 20/20 vision, it would still be virtually impossible for such a person to discern micro lettering at a distance. Cognitive speech and auditory processing are also important features when it comes to impressing humankind. 'A robot can listen to your heartbeat through a concrete wall, to your breathing underwater, and to your footsteps on solid ground.'[5] Human speech recognition patterns may take them slightly longer to master, especially given the innumerable intricacies and complexities of each language spoken on this planet.

'The average human has speech recognition down cold by age of three, and by adulthood has a recognition rate of about 99.2 percent and a 75,000-word vocabulary.'[6] Humans often speak too quickly, slur their words, mumble and use grammar incorrectly, and therefore androids may always struggle to interpret the lingo for the present. This is changing as artificial intelligences improve their analytical prowess of actions and consequences using concise, unemotional logic.

If androids are ever to truly interact with real people, communication and understanding is the robotic key. For a simple but valid example, I allude to an experimental 'robotic receptionist' that was trialled dealing with real people. Seemingly this experiment worked particularly well because 'she' behaved '… like a normal receptionist, sometimes politely answering questions and occasionally ignoring people, preferring instead to talk to her boyfriend [a new IBM mainframe]on the phone'.[7]

By virtue of an almost humanlike physical appearance and ultra-sophisticated senses, an android would have the inherent qualities necessary to potentially imitate human behaviour, and thus be reasonably interactive. However, looking and appearing to behave like humans has considerable limitations attached, particularly with respect to developing emotions, nurturing an intellect and, thereby, possessing a distinctive persona.

Androids will usually understand human behaviour but not necessarily the emotive reasons behind such behaviour. Will an android ever become angry, anxious, depressed, deceitful or experience sorrow? Can an android ever comprehend the subtle nuances when a human says one thing but means another? Will an android appreciate the conundrum when a human elects not to make a decision or choice, regardless of sufficient available information? Can their cognitive ability embrace feelings?

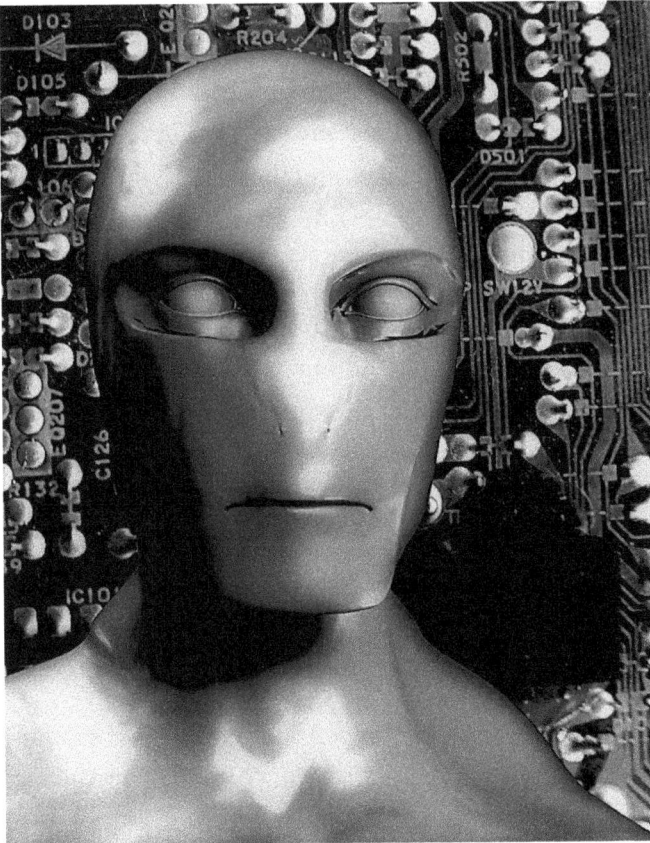

Figure 23: Developing the Humanoid

(© Shutterstock)

ROBOT AWAKENING

Humans communicate with each other using a vast array of subtle or pronounced processes – such as eye contact, facial expressions, speech, body movement and many more not as easily recognisable – in various combinations. This is where there may be a potential conflict. Does an android possess the essential comprehension to embrace the myriad of human behaviours?

Both humans and androids commence their existence with pre-programmed protocols. Children are born with distinctive inherited traits and the unique biological DNA composition that distinguishes every individual of the species. As they grow and mature, their behavioural 'protocols' are influenced and modified by their life experiences and various mentors, such as their parents, siblings, friends, colleagues and employers. Their freedoms to choose and act are moulded and developed by many external influences.

Conversely, androids may come into conflict with their programming in certain circumstances, unless they possess the freedom to self-regulate based upon their external influences and the capacity to learn from such experiences. Their capability to make unbiased, responsible decisions should not bring them into internal conflict with their pre-programming.

Children learn from a very early age to speak, readily recognise facial features and comprehend language without any conscious effort.[8] Ask any six-year-old or seven-year-old about the difference between right and wrong, and you would receive a very clear response. Now pose the same question to an intelligent android, but not before clarifying the following subtlety – that right can be wrong sometimes, or wrong can be right. In other words, sometimes being right about a matter is actually wrong (and overly smart compared to others), and deliberately being wrong is the right choice on the specific occasion (to avoid hurting someone's feelings for example).

An android can mimic human speech and even interpret languages, but can it distinguish between right and wrong? It could recognise through its programming that certain principles and responsibilities represent the right way, and everything else would be the wrong way by default.

However, can an android possess a moral compass? A simple example would be a grossly overloaded rowboat adrift on the ocean in rough seas. One passenger courageously elects to jump overboard so that the boat will not sink, thereby saving the remaining passengers. Most humans would try to retrieve the heroic passenger to prevent the person from drowning, although the boat would then become overloaded again. This would be the right way. An android may logically surmise that the right way is not to save the passenger, as the lives of the remaining people would then be at greater risk from the sinking of the boat. These are two different viewpoints on 'the right way' to salvage a dangerous situation.

Letting the passenger drown would conflict with Asimov's 1st law of robotics whereby 'a robot shall not … through inaction, allow a human being come to harm', and therefore the android may alternatively jump overboard in a valiant effort to retrieve the drowning passenger.

It is highly doubtful that futuristic androids will stay afloat in seawater, and thereby this alternative right decision becomes the wrong way and the android sinks. If it sounds confusing, imagine what an unfortunate dilemma faces the poor android. Stay in the rowboat and become ostracised by the remaining passengers for doing nothing at all to save a drowning human, or jump overboard and sink unceremoniously beneath the waves?

As well as having superior mental/intelligence skills, it is likely that future androids will eventually possess extraordinary physical skills. To an extent, this is already becoming clearly evident in their agility and versatility to function in a diverse range of environments hostile to humans. From space stations to chemical plants, from dangerous subterranean excavations to lifting extreme weights, using androids provides an increasingly safer alternative for humankind.

If androids are to effectively imitate humans in order to successfully perform their assigned duties, it will probably be essential that their self-learning processes are likened to those of people. Unfortunately, this may not always produce the optimum outcome in real life. In every batch of good eggs, there might just be a bad egg. Nobody ever said that learning was going to be an easy process. After all, 'Change is the end result of all true learning …', as quoted by American author Dr. Felice Leonardo (Leo) Bus-

ROBOT AWAKENING

caglia.

Futuroids

It is nature's way that life expands to fill any vacuum.
But when human minds become circuits
and human flesh becomes steel,
what will that new life reveal? [1]

In Our Own Image, *The Outer Limits (1998)*

Futuroid is another robotic term for individuality. If one needs to distinguish how 'future intelligent robots' may evolve, the following commentary may provide some useful guidance:

'… humanoid robots are good for studying the effectiveness of having a human-like body, and androids … for seeking the general nature of humans, studies using Germinoid [android robot duplicate of an existing person] focus on investigating the nature of individuality … to examine personal aspects (personality traits), trace their origins and implementing them into robots…' [2]

It is this complexity of combined personal aspects that makes an individual of each of us and, as a result, is a fundamental consideration in the development of advanced futuroids that will exhibit specific elements of human nature. They would not only resemble us closely in appearance, but behave in a similar fashion, within limitations.

So how shall such advanced androids acquire 'a human presence', and will it be as we expect? One approach could be by learning the innumerable and often complex traits and mannerisms of real people, and interacting in similar ways, thus gaining traceable knowledge. This is not necessarily an easy task given the subtle differences even in humans between talking and listening simultaneously, for example. The obvious sophistication of this inherent ability in humans may not be so readily acquired by autonomous AI, which are configured differently to organic life forms. In simplest terms, AI will receive and transmit in a totally different fashion to us.

They are engineered to be functional, efficient and thorough. Do robots really need emotions as well? They will if designed to successfully socially interact with humans. They will also require a distinctive name to confirm their individuality, and not simply a manufacturer's serial number.

One way to embrace this process is to think of futuroids as possessing the same 'conscious and unconscious recognition processes as humans' – to act and deal with people as if they were also human. By thinking like us, their responses should also conceivably be similar to those of real people. Of course, this will be greatly dependent upon many factors, particularly their perception capability to deal with the complexity of the human language, physical body and facial gestures and most importantly, the social and emotional behavioural traits of people.[3]

I have little doubt that the futuroid will eventually develop a superior intellect to humans as the cognitive ability of artificial intelligence evolves with time. The ongoing uptake of immense volumes of knowledge in a suitable intelligence database that is constantly expanding should in turn generate considerably improved intellect. What I am unsure about, with all this knowledge and learning, is how a futuroid will develop in the more subtle areas of human nature.

Without overly complicating matters, consider the essential human qualities of imagination, intuition (also known as 'the sixth sense') and emotions. I select these three distinctly different and often intangible human attributes intentionally to highlight some important differences between humans and AI. When a person is endowed with a highly imaginative mind, they are sometimes labelled as a dreamer or even a fanciful time-waster. Unless the human imagination is utilised to generate productive finite outcomes, whether it be for scientific research or aesthetic poetry, personal criticism may tend arise. Creative individuals who display original imagination are in many ways not always fully understood or appreciated by their peers.

Intelligent androids will probably evolve with a different type of creativity and conceptual output based upon their acquired knowledge. If truly autonomous, then futuroids will need to generate imaginative solutions and behaviours based upon their experiences. However, as the famous quote goes: 'Do androids dream [of electric sheep]? ...'[4] If it were at all plausible that AI could actually dream, would it be about circuits, diodes and algorithms, about analysing data, or would it possibly even be about other futuroids?

One 'dream' that may be possible, for example, is when a futuroid relatively indistinguishable in appearance from a real child, finds a human youngster – who might not realise that it is an android – for a friend. After all, imagination can be a marvellous gift.

Intuition can have several meanings, including natural instinct, foreboding of a future event, 'gut feeling' about a circumstance, or a perceptive sense that is inexplicable using the other five faculties ('sixth sense'). Intuition can be difficult to justify and does not require conscious reasoning, yet may ultimately prove to be correct. This ability is not always evident in many people, and yet can be very dominant in certain other individuals. In less learned ages, such an innate gift might have been regarded with suspicion and mistrust, and thought to be an extraordinary power beyond human reasoning. Superstitious folk even regarded such a sense as the means to determine the outcome of premonitions. In some respects, the human sixth sense is a gift of the mind that assists people in making decisions regardless of logic.

This is where any android with artificial intelligence may struggle, because L-O-G-I-C is the staple component of the futuroid. Without clear, concise and thorough logic to guide the decision-making processes, the futuroid would have some difficulty, and may even incur system shutdown by relying upon illogical factors to deliberate.

As a simple illustration, imagine a difficult steeplechase horse race with

20 starters over a long distance of 20 furlongs. There are many pertinent factors to consider to select the potential outright winner. These include each horse's weight, physique, temperament and previous competitive performances, the successful riding experience of each jockey, track condition and distance of race, prevailing weather and betting odds. AI would intensively interrogate all available pre-race data and possible outcomes to logically determine the most probable potential winner. The computations involve intricate statistical risks and various mathematical approaches. The AI scientifically selects the obvious favourite horse to win. A person with intuition would select one particular horse based only upon a gut feeling or basic instinct. The eventual winner is a horse with limited previous success and rated poorly in the betting, yet is correctly selected by the human despite the best scrutiny of the AI.

The reason (if one was needed) is that in a horse race with so many complex variables in play, even computer logic would struggle to determine the winner, and intuition can be a very powerful personal ally. 'Belief is the basis of intuitive behavior.'[5]

Intuition may never be reproducible in futuroids due to the intangibility of this sense. Humans tend to be emotional beings, often overriding sensible logic, and this defines their individuality. The same may not be said for AI devoid of emotions/feelings. For a futuroid to exhibit individuality compatible with *Homo sapiens*, it must have a sensitivity towards others and a due respect for their emotional values.

An old adage defining the uniqueness of humans is that 'the mind dispenses logic, but it is the heart that influences emotions'. The conflict between the two processes and the resultant decisions are what makes us human and individuals. Futuroids may possess self-programming to learn and develop sophisticated sub-routines similar to emotions, but it would be an arduous and prolonged journey.

One internationally acclaimed robotics researcher and creator Professor Hiroshi Ishiguro expects that an android that may eventually be completely indistinguishable from humans in appearance and behaviour will take about a hundred years to be realised.[6]

To appreciate how that robotic journey may evolve, I refer to the term 'empathy intelligence', particularly in the applied use of androids staffing the various healthcare industries for the aged, infirmed, those undergoing major disability rehabilitation and those suffering chronic illness. These institutions dispense long-term care, including not only assisting the elderly/disabled, but also providing daily companionship along the way. In this context, the word 'care' is defined as '… attention, warmth, kindness, reciprocity, empathy, and helpfulness'.[7] This is not to be confused with the provision of robotic services that are expected of such AI, such as their supportive physical functions.

Some of the recent technological and research developments in this area include the following data collection and assistance units: The human-lifting nursing robot RIBA II (Robot for Interactive Body Assistance); the in-home KSERA (Knowledgeable Service Robots for Aging), that learns by observation monitoring of daily behaviour and health of elderly occupiers and those with age-related illnesses, respiratory disease in particular; and HOBBIT (The Mutual Care Robot), that provides social assistance for elderly in fall prevention and detection around the home.[8]

To understand what makes an empathic robot/android, first it is vital to clarify what is known about human-robot interaction. In order to instill emotional skills in an android, there are three acknowledged abilities needed: the android must recognise the emotions of others [humans], be able to express them [in dealing with humans], and – the ultimate challenge – must possess emotions.[9] Recognition of the entire diversity of human emotions is probably the key to solving the overall concept for emphatic androids, and there is still a long way to go.

For an android to become emphatic, it needs to 'interpret' the emotions of humans during the interaction process, and respond accordingly to these cues by adapting its behaviour. Comprehending emotional signals can be complex, sometimes misleading and often confusing. Humans may say one thing and mean something totally different. Facial expressions and body/limb gestures can transmit contrasting messages. Contradiction equals confusion and emotional overload for an android.

Figure 24: Emotions of the Futuroid

(© Shutterstock)

The emotion 'minefield' of excessive signals distracts the android from intelligent communication, whilst expressing insufficient emotions creates doubt about the most suitable response. Another complexity to consider is that an android needs to produce realistic behaviour relative to humans, and not erratic, uncertain or fanciful responses. This would translate as disrespectful.

In the simplest terms, emotions need to be instilled as follows: a trigger stimulus occurs, the perceived emotion is affectively appraised, a response behaviour is assigned and then displayed interactively.[10] This truly robotic solution to a human condition is known as a mirror mechanism, whereby the android delivers a response dependent upon what it perceives.[11] The autonomous futuroids in years to come will conceivably possess simulated 'robotic emotions'. These will enable them to function at least as intelligently and robustly as humans, subject to the limitations inherent in the futuroids becoming able to successfully interact and learn from these experiences. For the present however, the process still remains in the research phase.

Assuming that the futuroid becomes as human-like as us, how will we positively determine it is actually artificial intelligence? There are some significant indicators that may help. Does it smell like a brand new soccer ball? Do its eyes follow your every movement with infinite precision? Does it respond to anything humorous in a sombre impassive tone? Most importantly, are you overcome with a creepy feeling that the near-perfect human is actually a near-perfect futuroid?[12] If the answer is affirmative to all questions, you have exposed the robotic truth.

The highly rational but eccentric cybernetic humanoid John Cavil, aka Cylon Model Number One, from the re-imagined USA television series *Battlestar Galatica* as shown on Sci-Fi Channel, expresses the right sentiment when he exclaims:

> 'I don't want to be human. I want to see gamma rays. I want to hear X-rays and I want to smell dark matter … feel the wind of a supernova flowing over me. I'm a machine, and I can know much more. I can experience so much more, but I'm trapped in this absurd body... '[13]

Antithesis

Oiled, with tube bones cut from bronze and sunk in gelatin, the robots lay.
In coffins for the not dead and not alive, in planked boxes ...
There was a smell of lubrication and lathed brass ... a silence of the tomb yard ... Named but unnamed, borrowing from humans everything but humanity... in a death that was not even a death, for there had never been a life.[1]

Ray Bradbury, *The Martian Chronicles (1950)*

The path of innovative research, development and evolution for advanced 'intelligent' robots is littered with superseded prototypes. The demise of these unique experimental bots is ultimately based upon sound scientific principles, most notably, replacement with superior or enhanced technology. However, one fascinating aspect to these decisions is the potential impact of 'infected' computer programming of AI.

Conceivably, even the most durable and sophisticated bots may still possibly be susceptible to the ravages of viral computer programs, 'sleeper' bugs/trojan horses/worms embedded in foreign software and other equally destructive software programs that may sometimes elude the most powerful anti-viral safety mechanisms. If ever advanced intelligent bots had one inherent weakness, it may be the malfunctioning side-effects of viral software, albeit unlikely. Strangely, these destructive influences may sometimes be the unfortunate result of unintended actions rather than any intentional sabotage. Conflicting consequences and required responses may disrupt the logic processes to such an extent that the dedicated programming becomes dysfunctional.

Figure 25: Defective Programming
(© Shutterstock)

I have no doubt that there are countless design safeguards engineered into these scientific systems to preclude, isolate and effectively thwart viral software. Given the immense financial investment and valuable intellectual rights in such AI technology, the risks have to be limited by suitable preventative safeguards.

Thereby lies the conundrum. What if AI can actually develop malware [destructive or disruptive software] customised to permit viral software to thrive? Malware that surreptitiously avoids detection initially until it has subtly altered key software and then is expunged leaving no trace. Superior recognition anti-viral software also developed by AI may in turn resolve the problem. The software combatants effectively become AI adversaries.

The issue remains that such bots may always be susceptible in one way or another, and how will this affect their decision-making processes? It is also crucial to understand whether these decisions are the result of either altered/dysfunctional programming, or merely the product of sentient AI itself [self-aware, autonomous machines with human-level intelligence].

In the realms of science fiction as portrayed in novels, television and movies, the principal causes of AI malfunctioning are usually incorrect or deficient programming, conflicting values or logic by the artificial intelligence, or some external 'alien' force altering/controlling the programming. In reality, it would be rare for infected software to occur within them, given the layers of software safeguards.

What may be more theoretically possible is the introduction of altered values and ideas, either directly or indirectly, that modify how the AI behaves and resolves issues. In other words, a subtle transition from the established truths and logic to a darker, less precise set of values. Once corrupted, the 'modified' behaviour then becomes the benchmark, skewing future logical decisions by the AI. If it sounds ominous and too unsettling, it is after all only a theory. The most difficult 'infection' to combat is that which is transient, constantly changing and adapting to suit the prevailing circumstances, thereby effectively becoming increasingly arduous to isolate for removal.

In 'Stream of Consciousness' (1997) from The Outer Limits Sci-Fi TV series, the worldwide AI technology has advanced to such a sophisticated culture that most people undergo a neural implant to permit a direct connection for a continuous uninterrupted flow of thoughts and feelings from the web (known as 'The Stream'). This intelligence becomes the source of all knowledge and can be accessed instantly by everyone, resulting in the consequential discarding or elimination of all other suitable media. Unfortunately, The Stream's erroneous programming results in a relentless search for all possible information at any cost. Such is the omnipotence of The Stream that it directs humankind to find every single piece of information and any remaining books, it disposes of those who do not obey. People become enslaved and dispensable in the insatiable quest for additional information. The Stream will not be shut down.

The ultimate dilemma to acquiring all data is in the cost of acquiring it. The hero who eventually succeeds in disabling the AI is someone who relied upon natural learning processes gained through reading books and had no neural implant. Originally treated by the population with little respect for not being connected to The Stream's interface, the hero remains the sole survivor of the human species still possessing knowledge. Without any further access to The Stream, the remainder are left in a residual childlike mental state. The re-education of *Homo sapiens* will need to resume from the beginning, using books instead.[2]

'You don't have to burn books to destroy a culture. Just get people to stop reading them.'[3]

If too much control by enhanced artificial intelligence creates obedience and servitude by humans, can the same be said of robotic self-development and autonomy? At what point does humankind pull the plug on AI before it indirectly enslaves them? Will stunting the learning processes of intelligent bots restrict their optimum potential? For this perplexing quandary, I turn to the matter of self-replicating bots. Any AI android capable of successfully manufacturing another exact duplicate android is self-replicating. Humans are capable of cloning living organisms, albeit under stringent ethical scrutiny and conditional upon prescribed clinical compliance. Will autonomous androids ultimately generate more androids?

Research in the domain of artificial self-replicating robotics has been underway in various forms since the late 1940s, originating with John van Neumann's conceptual self-replicating automata [universal constructor] and subsequent ongoing development of the logic of such machines.[4, 5]

Recent advances have accelerated this progress considerably, with researchers at Cornell University in New York successfully developing a modular 'robot' in 2005 that assembled itself in two and a half minutes. Comprising four cube-shaped separate modules ('molecubes') that each functioned independently, the experimental robot's construction clearly demonstrated self-reproduction was possible in a laboratory environment.[6] A self-assembling prototype microbot comprising cubes with no external moving parts ('M-blocks') has since been developed by other researchers in 2013, with the objective to eventually construct an extensive 'army/swarm of micro-bots'.[7]

These and other technological advancements in modular, self-reconfigurable robotic systems are really only the first early steps of considerably more scientific stages towards the process of robotic self-replication. Given these actual developments are not yet generating intelligent androids that are either replicants (an exact replica of a particular human) or indeed, self-replicating (an exact duplicate of themselves), I shall instead focus on the realms of fiction and what might be.

What would make an intelligent android of the future with artificial intelligence so potentially different to a human being? Certainly not physical appearance nor mental capability, but what about self-worth? Do androids care about their exterior appearance, their superior intellect or, for that matter, other fellow androids, like people care about themselves and friends, relatives and colleagues? In Philip Dick's famous fictional Sci-Fi book *Do Androids dream of Electric Sheep?* about fugitive replicant androids, the assumption is that they do not: '"An android", he said, "doesn't care what happens to another android. That's one of the indications we look for" [when interrogating them]'.[8]

However, there is one defining and salient difference that may ultimately distinguish humans from AI androids, and that is how we approach the end of our existence. For an android life form, it may well perceive this cessation of existence as an expiry, a permanent shutdown of all circuitry or a termination of functions. The outcome is the same. The antithesis of the birth of a human or the creation of an artificial life form is the demise of those beings. The difference lies in how we each view the final moment – a human about to die may exclaim 'I don't want to die!', whereas an intelligent android about to expire enquires 'what does it feel like?'..........................re-program please? '

Clanking, Squeaking & Rust

The only regret in my life is that I am not someone else.[1]

Woody Allen, epigraph, in Eric Lax,
Woody Allen and His Comedy (1975)

People, animals and artificial life forms, such as robots and androids, have individual peculiarities or foibles. In people, this may be evident as eccentric behaviour or various idiosyncrasies, such as unconventional or odd inexplicable actions. In animals, it may be more related to their natural behaviour that appears to us to be out-of-character for their species.

A humanoid mechanical robot of bygone years usually appeared to us as relatively fearsome in appearance and certainly as a manufactured being. Even in motion, the metallic joints and limbs tended to clank or squeak, much like a metal suit of Medieval armor. Their ability to walk or talk without appearing obviously 'robotic' with a stiff gait, awkward shuffling movements and a monotone voice only confirmed our impressions. However, once humans became accustomed to such artificial beings, the subsequent development of advanced androids almost indistinguishable from people (in some instances) created an entirely new perspective. Now, we had to decipher any subtle differences between such artificial intelligences and ourselves.

The clanking and squeaking of metallic body parts has long since gone and in its place, androids with sophisticated circuitry and elaborate facial and body features have evolved. So what distinctive foibles could be attributed to these New Age androids?

The most obvious peculiarity to me is their process of logic analysis. As humans cannot possibly process vast amounts of data as efficiently nor ascertain logical outcomes as clinically, we may flavour our logic with emotions. People tend to orientate their logic towards their particular sentiments at the time. Androids, however, do not have such a problem. Clear, concise and direct decisions are their priority. Reasoning is based upon sturdy factual information devoid of emotion, which merely confuses them. Take the following illogical witticism from American Comedian Woody Allen, for example:

'All men are mortal.
Socrates was mortal.
Therefore, all men are Socrates.'[2]

Most people would ask 'who is Socrates?' whereas an AI android would probably enquire 'why are all men mortal?' Logic is fine as long as you do not take it too seriously.

In any verbal communication between two beings, there is speaking and there is listening. Some people can do both simultaneously, whilst others are good at either talking or listening. Androids may have some difficulty if being asked two questions concurrently or if the replies require several possible answers. Humans appear very adept at talking, often saying something but meaning the opposite. What may be humorous to a person is not so clever to an android who dissects the information much like filleting a fish. For this example, I refer to the following:

How many robots does it take to screw in a light bulb?
Three – one to hold the bulb and two to turn the ladder.

A human would probably take this as a cynical quip, whereas an AI android would wonder why robots need to change light bulbs at all. On the subject of speech and clear enunciation of words, try speaking to a robotic telephone operator who constantly hears somewhat different words than those spoken by the caller. A simple yes or no usually suffices to satisfy the operator.

One particular disadvantage of being an AI android is being asked to share a joke with humans, as it is ALWAYS fraught with logic. One of my favourites is the following:

> A robot walks into a bar, orders a drink, and lays down some cash.
> Bartender says, 'Hey, we don't serve robots!'
> And the robot says, 'Oh, but someday you will.'
> The robot then makes a noise similar to someone gargling with water (akin to robotic laughter), but the humans remain silent and wondering why they asked the robot in the first place. Everybody knows that New Age androids cannot rust due to their high-tech componentry and super alloy/synthetic fabrication, but do androids know it? The rule is never to tell any joke pertaining to rust or metal corrosion, just to be safe.

Another distinctive linguistic oddity about some AI androids is their ability to communicate in mathematical formulae rather than words. I am no expert in the higher level mathematics, but refuse to communicate with them using another computer to comprehend the details. It defeats the entire purpose of intelligible speech. I would also classify an AI android's irritating ability of always being right about a topic or subject matter as another foible worth noting. I can only presuppose that being right is based upon their superior intelligence and infallible decision-making L-O-G-I-C. The solution is not to ask the question if you do not like the answer.

AI androids have progressed markedly in the past decade, to the point that their physical capabilities surpass many human endeavours. Best not to engage an android in an arm wrestle to avoid any potential limb injury. A competitive foot race is also out of the question, unless you blindfold the android and, ideally, select a treacherous course through swampy/marsh land at night.

Most humans enjoy music – either listening, composing, playing instruments or dancing to it. If an AI android can exhibit all of these skills, particularly synchronized dancing with another android, then it can definitely be my friend. One of the hardest chores for a person is coordinating both feet to dance steps without becoming entangled. The adage 'having two left feet' does not apply to androids because dancing is about logical sequential stages, and androids tend to shuffle anyway – the safest way to avoid becoming entangled.

Another peculiarity about AI androids is how they embrace silence. If you do not ask it anything or expect any action on your behalf, it is most likely to stand perfectly motionless as if frozen in time. The android's gaze will become more of a glaring, penetrating stare. Is it contemplating the next question or request, or does it simply hibernate? What is going on inside that cybernetic artificial brain? Can I outlast this extended period of deathly silence? Then, as if on robotic queue, the android suddenly spouts 'Can I get you anything?' Thoughts of a quick galactic space voyage to the nearest star constellation or even the nearest galaxy occur to me, but I gracefully respond 'No thanks!'

Whilst on the topic of an android's facial expressions, there is definitely something unusual about their eyes, and in particular, how they tend to track your every movement or gesture. Furthermore, the eyes in a human being are a significant indicator of a person's mood and disposition. Android eyes are the complete opposite, and tend to signify a blank, almost hypnotic trance-like state. Some even describe them as glassy, glazed or dormant. Of course their vision is entirely different to those of people, and so they may perceive their surroundings more as groupings of pixels than finite objects. I cannot blame them for having a distant, almost confused look in their eyes after all.

Probably the greatest disadvantage that an AI android may ever experience in its functional existence is having to go fishing on a small boat. This requires exceptional skills and very specific expertise. It may be manually retrieving the boat's heavy and awkward anchor stuck in mud or in a coral reef, delicately attaching slivers of bait onto sharp fish hooks and hauling in monstrous deep sea behemoths, cleaning and cooking said sea behemoths, or navigating the boat safely back to dock. The challenges are not insurmountable, but require sensible water-proofing at the very least, and a keen sense of housekeeping to keep the vessel shipshape and tidy. I am unsure what salty seawater does to electronic circuits, but it cannot be good if the android falls overboard.

The most tangible oddity about androids is their very existence. Have we manufactured them in the identical image of ourselves to facilitate our acceptance of them, or is it something more furtive? If it is to make us feel more comfortable in relating to androids, why might they appear not to be like us? Ignoring the obvious physical similarities, they remain an artificial life form and consequently would never realistically experience the identical emotions, ethics and responses of people. Thereby lies the conundrum – like us but not like us.

'Androids were never meant to dream.' – Tim Lebbon, *Predator: Incursion* (The Rage War; 2015).[3]

The Great Unknown

The oldest and strongest emotion of mankind is fear, and the oldest and strongest kind of fear is fear of the unknown.

H.P. Lovecraft, *Supernatural Horror in Literature (1927)*

As the world's technological knowledge and scientific comprehension about artificial intelligence advances, there appears to be just as many questions raised as answers delivered. Will AI eventually develop beyond any autocratic control by humankind? Will this mean they will be able to self-replicate and construct life independently of all external assistance? If such life forms operate in conjunction with the human species, will they remain in their present role, providing critical services and support for us, or will they perhaps progress to an entirely autonomous society?

There appears to be many varied viewpoints postulated on these concepts or derivatives of these matters, without any clear outcomes. Organic life forms such as humans have created these synthetic beings and, although there remains a long way to go on this research journey, the early part of the 21st century has definitely seen an acceleration in the development of AI. In so many ways, the science fiction of the past is progressively becoming scientific fact of the present.

To address some of these concerns objectively, I examine recent popular robotic advancements as a suitable signpost for the future. Autonomous/

driverless automobiles are an inevitability for humans. Once all the legal ramifications and engineering design obstructions are effectively resolved, there should be a proliferation of such vehicles. Drivers will become silent passengers and the vehicle will do most of the work.

Of course, there is always an 'x-factor' in any major discovery, and I surmise that this is human unpredictability. This largely unknown factor is how and why people behave so differently on our roads – emotional stress, vehicular inexperience, traffic delays, impatience, road condition, inclement weather, call it what you wish. An exceptionally sophisticated AI system may be required to safely operate the vehicle and avoid a myriad of such potentially hazardous circumstances. The present human drivers of this world will certainly need to be re-trained to cope with simply being the passenger in such a robotic vehicle.

Medical developments in robotic surgery are increasing, ably directed by dedicated professional surgeons and other specialists. This union of technological precision and medical providers is definitely here to stay, as advancements in rapid disease detection and clinical removal/non-invasive procedures continue to evolve. If such proactive systems increase the longevity of human life and provide lifesaving relief for others, then robotics is the way of the future. On a more controversial scientific path is the subject of human cloning. I have little doubt that replicating intelligent androids will occur well before people progress too far along this highly sensitive avenue. Replicant 'futuroids' could probably deliver a far safer alternative for humankind, as long as pre-determined programming logic safeguards are implemented and cannot be corrupted or distorted.

One of the gravest concerns for those opposing the rise of the machines is not only the capability of certain research robots to learn efficiently and rapidly ('deep learning'), but also to share this knowledge with other intelligent robots ('teacher bots'). These fundamental steps to most advanced civilisations are the building blocks for any future robotic society. The learning and subsequent sharing of vital information, particularly through a global network, is already here, but has to date been developed by people.[1]

If the current revolution occurring in industrial robotics is any indication, it should not be long before teacher bots operate entire factories. With countries like China already embracing robotic technology-driven expansion to deliver improved manufacturing efficiencies, the scale of such innovative change could be enormous and irreversible in the next few years.[2]

So what is the concern about smart, artificially-intelligent machines able to share their knowledge with other like machines? Surely it would be a win-win situation for both humankind and their creations, and ideally, would deliver harmonious outcomes for our planet. As with some many other species, adaptability to change is imperative to survival. Adaptive robots may just be a technological extension to this primary rule of nature.

Such adaption may include such an unprecedented increase in robotic services worldwide that humans will probably come to expect improved living and working standards as the norm. A society of sophisticated service robots would be interconnected by their own 'intelligent' global network to deliver the optimum efficiency for people and businesses alike. The smart homes, factories and city facilities of the future are already on the robotic horizon, perhaps not yet clearly discernible to everyone.

How many new homes are now being built incorporating automated user-friendly systems for their occupants, such as integrated audio/touch sensors and programmable controllers? Factories are progressively installing impressive computer-controlled precision equipment or fully automated robotic facilities with minimal human intervention in manufacturing and in many other industries. The introduction of mobile autonomous drones to scope a myriad of previously inaccessible geographical locations, or perhaps highly hazardous locations, is well underway. So far, none of these examples should represent a concern for *Homo sapiens*.

From my perspective, the three greatest unknowns that may seriously impact on our lives in the future are almost interrelated. With vastly improved robotic personal services and superior medical advancements, most of us should potentially have an extended lifespan. We should live far longer than then our predecessors. An entire population living considerably longer presents a complex suite of socio-economic problems not able to be addressed in this book. Our work routines will not resemble current jobs, if indeed we still need to work at all.[3]

In an extensive robotic society, humans would probably only provide the higher level guidance and manipulation needed to ensure 'the system' continues to operate smoothly. Robots become providers and humankind merely the observers. So we should have an extended lifespan with plenty of leisure time, and would work in entirely different roles than in today's world, perhaps providing high level assistance and direction to the artificial intelligence providers.

The remaining unknown factor, which I tentatively name the 'y' factor, represents the ultimate conundrum and is the most difficult to predict. What if the AI we create and their self-generating systems no longer serve humankind? Whether by unintentional means, or simply as a result of self-serving robotics, the potential always remains that AI no longer needs our assistance or relies upon us at all. For this sombre conclusion, I resort to two somewhat different quotations for some subtle guidance.

In 1920, Czechoslovakian author and playwright Karel Čapek produced the influential science fiction play entitled *Rossumovi Univerzální Roboti* (abbreviated as R.U.R.), in which the word 'robot' was introduced to the English language, and the following sentiment expressed:

> 'Robots do not hold on to life. They can't. They have nothing to hold on with – no soul, no instinct. Grass has more will to live than they do.'[4]

In 2014, British novelist Matt Haig stated in an interview for *The Guardian* newspaper:

> 'Robots are great. I am saying that now, so that when a future civilization of robots takes us captive, they will search through *The Guardian* web archive and realise I said, "robots are great," and then they'll choose to save me.'[5]

Although the quotations are well over 90 years apart, the message remains the same. Robots are not human and we had better become accustomed to it. Historically, when they were made in our image and learned from us, it was usually for our benefit. To this end, I will the leave the matter of the 'y' factor unresolved, simply adding that only the future will show us the ultimate outcome for the human species in a robot-dominated society.

Epilogue

Poetry is an echo, asking a shadow dancer to be a partner.[1]

Carl Sandburg, 1923

I have selected salient poetic verses from three notable sources to distill the essence of this book, commencing with the famous and esteemed British poet, author and journalist Rudyard Kipling. His poem *The Secret of the Machines* (1911) provides us all with a poignant reminder of the nature of modern machinery that could so easily be applied to robotic entities.

We were taken from the ore-bed and the mine,
We were melted in the furnace and the pit—
We were cast and wrought and hammered to design,
We were cut and filed and tooled and gauged to fit.

Some water, coal, and oil is all we ask,
And a thousandth of an inch to give us play
And now, if you will set us to our task,
We will serve you four and twenty hours a day!

We can pull and haul and push and lift and drive,
We can print and plough and weave and heat and light,
We can run and race and swim and fly and dive,
We can see and hear and count and read and write! ...

But remember, please, the Law by which we live,
We are not built to comprehend a lie,
We can neither love nor pity nor forgive.
If you make a slip in handling us you die!

We are greater than the Peoples or the Kings—
Be humble, as you crawl beneath our rods!-
Our touch can alter all created things,
We are everything on earth—except The Gods!...

... for all our power and weight and size,
We are nothing more than children of your brain! [2]

The next partial verses were extracted from the poem *The Bells* (1849) by the famous American writer, poet, editor and literary critic Edgar Allan Poe. With a degree of enlightened imagination, the extracts certainly may have appeal to the realms of robotic life forms, especially given Poe's focus in this work on the darker mechanical elements of the topic.

What a tale their terror tells
Of Despair!
How they clang, and clash, and roar!
What a horror they outpour
On the bosom of the palpitating air!
Yet the ear it fully knows,
By the twanging,
And the clanging,
How the danger ebbs and flows:
Yet the ear distinctly tells,
In the jangling,
And the wrangling,
How the danger sinks and swells,...

… For every sound that floats
From the rust within their throats
Is a groan…

… They are neither man nor woman-
They are neither brute nor human-…[3]

Awakening can be a metamorphosis. For robots, it could be the ultimate transformation from subservient creations to autonomous life forms, and perhaps even sentient AI in future times. For this dramatic transition to occur, it must involve a technological 'awakening'. The famous American poet and teacher Henry Wadsworth Longfellow provides this impetus for me in his poem *Becalmed* (1882). For the reader, assume that it is a robot's 'thoughts' expressed throughout these verses:

Becalmed upon the sea of Thought,
Still unattained the land it sought,
My mind, with loosely-hanging sails,
Lies waiting the auspicious gales.

On either side, behind, before,
The ocean stretches like a floor, –
A level floor of amethyst,
Crowned by a golden dome of mist.

Blow, breath of inspiration, blow!
Shake and uplift this golden glow!
And fill the canvas of the mind
With wafts of thy celestial wind.

Blow, breath of song! until I feel
The straining sail, the lifting keel,
The life of the awakening sea,
Its motion and its mystery![4]

The ultimate challenge may then be fulfilled for such life forms.

References

CHAPTER 1: Replication of the Species

1. Shakespeare, *Hamlet*, Act I, Scene V, p.VIII, 1599-1602.

CHAPTER 2: History of Discovery

1. Automaton, http://www.dictionary.com/browse/automaton. Retrieved 4 September 2017.

2. Truitt, E.R., *Medieval Robots: Mechanism, Magic, Nature and Art*, 2015, pp.2-3.

3. Nelson, V., *The Secret Life of Puppets*, Harvard University Press, Cambridge, Massachusetts, 2001.

4. Russell, B. (ed.), *Robots: The 500-year quest to make machines human*, 2017, p.28.

5. Graetz, G., and Michaels, G., *Robots at Work*, 2015, p.2.

6. McKerrow, P., *Introduction to Robotics*, 1991, p.2.

7. Truitt, op.cit., p.4.

8. Russell, op.cit., pp.34-5.

9. Ibid., p.46.

10. Robotics Timeline, http://www.robotshop.com/media/files/PDF/time-line.pdf, Retrieved 5 September 2017.

11. Roberts, D., *Famous Robots and Cyborgs*, 2014, p.6.

12. Russell, op.cit.,p.46.

13. Robotics Timeline, op.cit.

14. Roberts, op.cit., p.7.

15. The Model Engineer and Light Machinery Review 59, No.1429, 20 September 1928, p.275.

16. Shalako, L.B.,'Mr Robot' in *Twisted Tongue Magazine*, Issue #16, June 2010.

17. Roberts, op.cit., p.8.

18. Ibid., p.7.

19. Telotte, J.P., *Robot Ecology and the Science Fiction Film*, 2016, p.16.

20. Wilson, D.H., *How To Survive a Robot Uprising: Tips on Defending Yourself Against the Coming Rebellion* 1st Ed., 2005, p.99.

21. Rosheim, Mark, *Robot Evolution: The Development of Anthrobotics*, 1994, pp.256-7.

22. Wabot-1, *Humanoid Robotics Institute (HRI)*, Waseda University, www.humanoid.waseda.ac.jp/booklet/kato_2.html, Accessed 8 September 2017.

23. 1993 -1997: Evolving to a Humanoid Robot that combines an Upper Body with Legs, *History of Honda's Robot Development*, http://www.world.honda.com/ASIMO/history/p1_p2_p3/ Accessed 11 September 2017.

24. Evolution of ASIMO, *The Honda Humanoid Robot*, http://www.world.honda.com/ASIMO/ technology/2011/index.html, Accessed 11 September 2017.

25. Honda Unveils All-new Asimo with Significant Advancements, *Honda Worldwide site www.world.honda.com*, 8 November 2011.

26. Mission to International Space Station – Robonaut, Accessed 11 Sep-

tember 2017.

CHAPTER 3: Science Fiction

1. De Ford, M.A., 'Foreword', in *Elsewhere, Elsewhen. Elsehow: Collected Stories*, 1st Ed., Walker & Company, New York, USA, 1971.

2. BBC, 'The Robots of Death', in *Doctor Who*, Season 14, Serial 5, Final episode originally aired 19 February 1977.

3. Roberts, D., 'Danger, Humans!' in *Famous Robots and Cyborgs: An Encyclopedia of Robots from TV, Film, Literature, Comics, Toys and More*, 2014, p.139.

4. Gunn, J., *Isaac Asimov: The Foundations of Science Fiction*, revised edition, 1996, p.41.

5. Ibid.,'Variations upon a Robot', p.47.

6. Asimov, I., 'Runaround', in *I,Robot,*(The Isaac Asimov Collection edition), 1950, p.40.

7. Wilson, op.cit., 2005, p.110.

8. Gunn, op.cit., p.117, p.91.

9. Ibid., p.93, p.107.

10. 'The Brain Center at Whipple's', *The Twilight Zone: The Original Series (1959-1964)*, Season 5, Episode 33, originally aired 15 May 1964, https://en.wikipedia.org/wiki/ The_Brain_Center_at_ Whipples.

CHAPTER 4: Industrial Bots

1. Morris, J., 'Employee of the Month', *Leicester Mercury* newspaper, Visual Humour publishing, Great Britain.

2. Rosenberg, J. M., *Dictionary of Artificial Intelligence and Robotics*, 1996, p.88.

3. Japan External Trade Organisation, *Industrial Robots, Now In Japan Series, No.32/1981*, 1981, pp.2-5.

4. Royakkers, L., and Van Est, R., *Just Ordinary Robots: Automation from Love to War*, 2016, p.4.

5. Duffy, J., ' Impact of Robotics in Manufacturing, Agriculture and Min-

ing Within The USA', in *Vol 1:Robots in Australia's future*, Conference Proceedings, Perth, Western Australia, 13-16 May 1986, p.2.

6. Japan External Trade Organisation, op. cit., p.41.

7. Ibid., p.35.

8. Asimov, op.cit., p.40.

9. Japan External Trade Organisation, op. cit., p.2.

10. Lovecraft, H.P., 'Chapter I. Introduction' in *Supernatural Horror in Literature, The Recluse Magazine*, 1927.

CHAPTER 5: Robot or Android

1. Royakkers and Van Est, op. cit., pp.10-11, p.44.

2. Wikipedia encyclopedia, https://en.wikipedia.org/wiki/Android, Accessed 1 September 2017.

3. MacDorman, K.F., and Ishiguro, H., 'Towards Social Mechanisms of Android Science', in Social Behaviour and Communication in Biological and Artificial Systems, *Interaction Studies Journal*, Volume 7, Issue 2, 2006, p.289.

4. Russell, op.cit., p.87.

5. Weiner, N., *Cybernetics*, or *Control and Communication in the Animal and the Machine*, 1948.

6. Royakkers and Van Est, op. cit., p.8.

7. BBC, 'The Caves of Androzani' *Doctor Who*, Season 21, Serial 6, Final episode originally aired 16 March 1984.

8. Cameron, K., 'Doctor Who Top 10', in *The Telegraph*, Issue 16 September 2009.

9. 'The Top 10 Doctor Who stories of all time', in *Doctor Who Magazine*, Issue 474, 21 June 2014.

10. The Doctor Who Transcripts – The Caves of Androzani, http://www.chakoteya.net/DoctorWho/21-6.htm. Retrieved 1 September 2017.

11. Roberts, op.cit., pp.81-2.

12. CBBC, *The Sarah Jane Adventures*, Sci-Fi Television Series 1-5, BBC Cymru Wales, 2007-2011.

13. *Shadow Dexterous Hand*, https://www.shadowrobot.com/products/dexterous-hand. Accessed 2 September 2017.

14. 'CYGAN – Dr Fiorito's Giant Electronic Robot', in *Modeling Review Magazine* (Italy), 1957.

15. British Pathé, Gentle Giant (aka Robot) film clip, 1958.

16. Russell, op.cit., 'Cygan, the mechanical man', p.84.

CHAPTER 6: The Complete Robot

1. Rosenberg, op.cit., p.10.

2. Caudill, M., *In Our Own Image: Building an Artificial Person*, 1992, p.8.

3. Russell, op.cit., p.163.

4. Ibid., p.158, p.161.

5. Hands On:FURo-i Home Personal Robot, http://www.roboticstrends.com/article/hands_on_furo_i_home_personal_robot. Retrieved 29 August 2017.

6. Marchionni, L., et al., 'REEM Service Robot: How may I help you?' in *Natural and Artificial Models in Computation and Biology, 5th International Work-Conference on the Interplay Between Natural and Artificial Computation (IWINAC 2013), Proceedings*, Part 1, Spain, 2013, pp.122-129.

7. Roberts, op.cit., p.3.

CHAPTER 7: Fembots

1. Belli, J., and Campoverde, J., *Gynoids: The Impact of Female Robots in Real Life*, ENG 2420, Science Fiction, 2016, p.2, http://openlab.city-tech.cuny.edu. Retrieved 13 August 2017.

2. Rosenberg, op.cit., p.8.

3. Liddell, H.G., Scott, R., Jones, H.S., and McKenzie, R., *A Greek and English Lexicon*, 1940, p.55.

4. 'The Lonely', *The Twilight Zone: The Original Series* (1959-1964), Season 1, Episode 7, originally aired 13 November 1959, https://en.wiki-

pedia.org/wiki/ The_Lonely_(The_Twilight_Zone).

5. Levin, I., *The Stepford Wives*, 1st Ed., 1972.

6. 'Bits of Love', *The Outer Limits* Canadian-American TV Series, Episode 1, Season 3, originally aired 19 January 1997.

7. Eliot, T.S., 'The Hollow Men', in *Poems 1909 -1925*, pp.127-8.

CHAPTER 8: Toy Robots

1. Bunte, J., Halleman, D., and Mueller,H., *Vintage Toys:Robots and Space Toys*, 1999, p.10.

2. Roberts, op.cit., p.9.

3. Ibid., p.20.

4. Emchowicz, A., and Nunneley,P., *Future Toys: robots,astronauts,spaceships, ray guns*, 1999, pp. 33-55.

5. Bunte et al, op.cit., p.104.

6. Ibid., p.46.

7. Ibid., p.56-7.

8. Ibid., p.36.

9. Ibid., p.152.

10. Ibid., p.110.

11. Lost in Space, https://en.wikipedia.org/wiki/Lost_in_Space. Retrieved 20 August 2017.

12. Bunte et al, op.cit., pp.162-4.

CHAPTER 9: A Child's Perspective

1. Wagoner, P.V., and Smith, S. (Illustrator), *Could You Hug A Cactus?* 2016, pp.8-9.

2. Adams, D., *The Hitchhiker's Guide to the Galaxy*, 1st Ed., 1979, p.63, p.91.

3. Ibid., p.142.

4. Adams, D., *The Restaurant at the End of the Universe: Volume 2 of The Hitchhiker's Guide to the Galaxy*, 1st Ed., 1980, p.96.

5. Adams, D., *The Hitchhiker's Guide to the Galaxy*, p.69.

6. Ibid., p.94.

7. Ibid., p.69.

8. Ibid., p.103.

9. Matronic, A., 'The Mighty Atom/Astro Boy', in *Robot Universe: Legendary Automatons And Androids From The Ancient World To The Distant Future*, 2015, p.96.

10. 'ANIME', *Merriam-Webster Online Dictionary*, 2011. https://www. merriam-webster.com/dictionary/anime. Accessed 15 September 2017.

11. 'Astro Boy', https://en.wikipedia.org/wiki/Astro_Boy. Retrieved 16 September 2017.

12. 'The Jetsons', https://en.wikipedia.org/wiki/The_Jetsons.

13. 'The Vacation', *The Jetsons*, Episode 30, Season 2, originally aired 7 November 1985.

14. 'The Swiss Family Jetson', *The Jetsons*, Episode 22, Season 2, originally aired 23 October 1985.

15. 'Rosie's Boyfriend', *The Jetsons*, Episode 8, Season 1, originally aired 11 November 1962.

16. 'Rosie Come Home', *The Jetsons* Episode 2, Season 2, originally aired 17 September 1985.

17. 'Rip-Off Rosie', *The Jetsons* Episode 23, Season 2, originally aired 24 October 1985.

18. 'Robot's Revenge', *The Jetsons* Episode 36, Season 2, originally aired 20 November 1985.

CHAPTER 10: Robot Society

1 Yu, Chong San, 'New Perspective of Robotics – Why Not Robot Society?', in *Vol 2: Conference Proceedings of Robots in Australia's future*, Perth, Western Australia, 13-16 May 1986, pp.23-5.

2. Caudill, op. cit., p.5, p.212.

3. Kurzweil, R., *The Age of Spiritual Machines When Computers Exceed Human Intelligence*, 1999.

4. 'Resurrection', *The Outer Limits* Canadian-American TV Series, Episode 2, Season 2, originally aired 14 January 1996.

5. 'The Tipping Point', *The Outer Limits* Canadian-American TV Series, Episode 19, Season 7, originally aired 15 September 2001.

CHAPTER 11: Rogue Robots

1. Wikipedia, *Chatbot*, https://en.wipedia.org/wiki/Chatbot. Retrieved 12 September 2017.

2. Mauldin, M., 'ChatterBots, TinyMuds, and the Turing Test: Entering the Loebner Prize Competition', in Believable Agents, *Vol 1:Proceedings of the Twelfth National Conference on Artificial Intelligence*, 31 July - August 1994, Washington, USA, pp.16-21. Retrieved 20 September 2017.

3. Oppenheim, M., 'Chinese chatbots deleted after criticising the ruling Communist Party', in *The Independent*, Issue 3 August 2017. http://www.independent.co.uk/news/world/asia/china-chatbots-communist-party-ruling-critics-peoples-tencent-babyq-little-bing-a7875601.html. Retrieved 12 September 2017.

4. Clark, B., 'Facebook's AI accidentally created its own language', in *businessinsider.com*, http://www.businessinsider.com/facebook-chat-bots-created-their-own-language-2017-6?IR=T. Retrieved 12 September 2017.

5. 'European Parliament calls for legislative framework governing creation, use and taxation of robots and artificial intelligence', EY Global Tax Alert, 9 February 2017. http://www.ey.com /Publication/vwLUAssets/ European_Parliament_calls_for_legislative_ framework_governing_creation_use_and_taxation_of_robots_and_artificial_intelligence.

6. Orseau, L., and Armstrong, S., 'Safely Interruptible Agents', in *Uncertainty in Artificial Intelligence (UAI)* - Proceedings of Thirty-Second International Conference, Jersey City, New Jersey, USA, 25-29 June 2016, pp.557-566.

7. Monstrousness, *Robots don't cry*, 2014, https://monstrouspoetry.wordpress.com/tag/ poetry/.

CHAPTER 12: Robo Poetry

1. Clark, B., *'Facebook's AI accidentally created its own language'*, Facebook Artificial Intelligence Research (FAIR) Laboratory, https://thenextweb.com/artificial-intelligence/2017/06/19/facebooks-ai-accidentally-created-its-own-language/. 19 June 2017.

2. Feinberg, J., 'The Idea of a Free Man', in *Educational Judgements: Papers in the Philosophy of Education*, 1973, p.161.

3. Dvorsky, G., *How Artificial Superintelligence Will Give Birth To Itself*, https://www. gizmodo . com.au/2016/06/how-artificial-superintelligence-will-give-birth-to-itself/. Retrieved 25 September 2017.

4. Wong, S., Google Translate AI, *New Scientist*, https://www.newscientist.com/article/ 2114748-google-translate-ai-invents-its-own-language-to-translate-with/ Retrieved 21 September 2017.

5. 'Google's AI has written some amazingly mournful poetry', http://www.wired.co.uk /article/google-artificial-intelligence-poetry. Retrieved 28 September 2017.

6. Bowman, S., Vilnis, L., Vinyals et al., 'Generating Sentences from a Continuous Space', arXiv: 1511.06349v4, *Proceedings of The 20th SIGNLL Conference on Computational Natural Language Learning*, Berlin, Germany, 7-12 August 2016, Table 12, p.12.

7. Poet Destroyer A, https://www.poetrysoup.com/poem/robot_216378. Posted 3 April 2010.

CHAPTER 13: Human or Android

1. Bao, Z., 'Organic Materials-Based Flexible Electronic Sensors', in *Frontiers in Chemistry Seminar: Frontiers in Organic Materials for Information Processing, Energy and Sensors*, Annual Meeting, American Association For The Advancement Of Science (AAAS), Washington, USA, 17-21 February 2011.

2. Xiandi Wang, Lin Dong, Hanlu Zhang, Ruomeng Yu, Caofeng Pan, and Zhong Lin Wang, 'Recent Progress in Electronic Skin', in *Advanced Science Vol 2*, Issue 10, 15 October 2015, pp.17-8. http://www.nanoscience.gatech.edu/paper/2015/15_AS_01.pdf.

3. Innovations/Technology, http://www.hansonrobotics.com/about/innovations-technology/. Accessed 2 October 2017.

4. Takanishi Laboratory, Waseda University,'Latest bipedal walking robot that can show various emotions', https://www.waseda.jp/top/en-news/13858, Retrieved 1 December 2017.

5. Wilson, op.cit., p.62.

6. Ibid., p.85.

7. Ibid., p.80.

8. Ibid., p.15.

CHAPTER 14: Futuroids

1. 'In Our Own Image', *The Outer Limits* Canadian-American TV Series, Episode 26, Season 4, originally aired 18 December 1998.

2. Ogawa, K., and Ishiguro, H., Osaka University, '*Android Robots as In-between Beings*', in *Robots and Art: Exploring an Unlikely Symbiosis*, Springer Singapore, 2016, p.328.

3. Ibid., pp.331-3.

4. Dick, P., *Do Androids dream of Electric Sheep?*, 1968, p.182.

5. Nadel, L., *Dr. Laurie Nadel's Sixth Sense: Unlocking Your Ultimate Mind Power*, 2006.

6. Royakkers and Van Est, op.cit., p.64.

7. Ibid., p.91.

8. Ibid., pp.101-5.

9. Picard, R.W., *Affective Computing*, M.I.T Media Laboratory Perceptual Computing Section Technical Report No. 321, 1997, p.1.

10. Breazeal, C., and Brooks, R., *Robot Emotion: A Functional Perspective*, 2004, pp.16-7.

11. Breazeal, C., Takanishi, A., and Kobayashi, T., *Social robots that interact with people*, in Springer Handbook of Robotics, B. Siciliano and O. Khatib (eds.), 2008, Chapter LXXI.

12. Wilson, op.cit., pp. 81-2.

13. John Cavil aka Cylon Model Number One, 'No Exit', in *Battlestar*

Galatica, Episode 17, Season 4, originally aired 13 February 2009.

CHAPTER 15: Antithesis

1. Bradbury, R., 'Usher II' (April 2005/2036), in *The Martian Chronicles*,1st Ed., 1950.

2. 'Stream of Consciousness', *The Outer Limits* Canadian-American TV Series, Episode 5, Season 3, originally aired 7 February 1997.

3. American Writer Ray Bradbury,"Bradbury Still Believes in the Heat of 'Fahrenheit 451'", Interview by Misha Berson, in *The Seattle Times*, Issue 12 March 1993.

4. Stevens, W., *Self-Replication, Construction and Computation*, PhD Thesis, Chapter 2, The Open University, 28 October 2009. 'History of Self-Replicating Machines' http://www.srm.org.uk/introduction2.html.

5. Penrose, L., 'Self-Reproducing Machines', in *Scientific American Journal*, Vol 200, Issue 6, June 1959, p.105.

6. Steele.W., 'Researchers build a robot that can reproduce', in *Cornell Chronicle*, Cornell University, NY, http://news.cornell.edu/stories/2005/05/researchers-build-robot-can-reproduce 25 May 2005.

7. Romanishin J.W., Gilpin, K., and Rus, D., *M-blocks: momentum-driven, magnetic modular robots*, in 2013 IEEE/RSJ International Conference on Intelligent Robots and Systems, Tokyo, November 2013, pp. 4288–95.

8. Dick, op.cit., p.99.

CHAPTER 16: Clanking, Squeaking & Rust

1. Rawson, H., and Miner, M., *The Oxford Dictionary of American Quotations*, 2nd Ed., 2006, p.576.

2. Wikipedia, *Love and Death*, Woody Allen (Director and Writer), Jack Rollins and Charles H. Joffe Productions, USA, June 1975. https://en.wikiquote.org/wiki/Woody_Allen# Love_and_Death_.281975.29. Accessed 18 October 2017.

3. Lebbon, T., 'Liliya - Testimony', in *Predator: Incursion (The Rage War; Book 1)*, 25 September 2015, Chp.7.

CHAPTER 17: The Great Unknown

1. Knight, W., 'Smarter learning', in *5 Robot Trends to look for in 2016*, MIT Technology Review, 1 January, 2016, https://www.technologyreview.com/s/545056/5-robot-trends-to-watch-for-in-2016/, Accessed 23 October 2017.

2. Ibid., 'China's robot revolution'.

3. Marr, B., '1, 2', in *The 5 Most Worrying Technology Trends For 2017 and Beyond*, Forbes.com. 23 December 2016, https://www.forbes.com/sites/bernardmarr/2016/12/23/the-5-most-worrying-technology-trends-for-2017-and-beyond/, Accessed 1 November 2017.

4. Čapek, K., Rossum's Universal Robots play, Act 1, 1920, https://en.wikiquote.org/wiki/ Karel_%C4%8Capek, Accessed 2 November 2017.

5. Haig, M., 'Matt Haig's top ten robots', in *Children's Books*, The Guardian, https://www.theguardian.com/childrens-books-site/2014/apr/10/matt-haig-top-10-robots,10 April 2014. Retrieved 2 November 2017.

CHAPTER 18 Epilogue

1. Sandburg, C., 'Poetry Considered', in *The Atlantic Monthly*, Vol. 131, 1923, p.342.

2. Eliot, T.S., *A Choice of Kipling's Verse Made by T.S. Eliot with an Essay on Rudyard Kipling*, 1st Edition, 1943, p.293.

3. Poe, E.A., 'Poe's Last Poem', in *New York Tribune*, Issue 20 October, 1849, p.1.

4. Longfellow, H.W., *In The Harbor*, *Ultima Thule.* – Part II., 1st Ed., 1882, pp.9 -10.

Bibliography

Acosta, Bremer, *Blood of Other Worlds*, 2nd Ed., Bizarre Warp Press, USA, 2016.

Adams, Douglas, *The Hitchhiker's Guide to the Galaxy,* 1st Ed., Pan Books, United Kingdom, 1979.

Adams, Douglas, *The Restaurant at the End of the Universe: Volume 2 of The Hitchhiker's Guide to the Galaxy*, 1st Ed., Pan Books, United Kingdom, 1980.

Allen, Irwin, *Lost in Space* Series, Irwin Allen Productions, CBS and 20th Century Fox Television, USA, 1965-68.

American Association For The Advancement Of Science, *Frontiers in Chemistry Seminar: Frontiers in Organic Materials for Information Processing, Energy and Sensors*, Annual Meeting, AAAS, Washington, USA, 17-21 February 2011.

Asimov, Isaac, *I, Robot*, Gnome Press, New York, 1950.

Asimov, Isaac., *The Caves of Steel*, Galaxy Magazine, Galaxy Publishing

Corporation, New York, Vol.7 Parts 1 -3, October -December, 1953.

Asimov, Isaac., *The Naked Sun*, Astounding Magazine, Vol LVIII, Nos.2 -4, Street & Smith Publications, New York, October-December, 1956.

Asimov, Isaac, *The Rest of the Robots*, 1st Ed., Doubleday, New York, 1964.

Association for the Advancement of Artificial Intelligence, *Volume 1: Proceedings of the Twelfth National Conference on Artificial Intelligence*, 31 July - August 1994, Washington, USA, AAAI Press, 1994.

Association for Computational Linguistics, *Proceedings of The 20th SIGNLL Conference on Computational Natural Language Learning*, Berlin, Germany, 7-12 August 2016, Publisher Association for Computational Linguistics, 2016.

Association for Uncertainty in Artificial Intelligence, *Uncertainty in Artificial Intelligence (UAI)-Proceedings of Thirty-Second International Conference*, Jersey City, New Jersey, USA, 25-29 June, 2016, AUAI Press, USA, 2016.

Australian Robot Association, *Volumes 1-2:Robots in Australia's future*, Conference Proceedings, Perth, Western Australia, 13-16 May 1986.

Belli, Jill, and Campoverde, Joselin, *Gynoids: The Impact of Female Robots in Real Life*, ENG 2420, Science Fiction, 2016.

Bostrom, Nick, and Ćircović, Milan (Eds.), *Global Catastrophic Risks*, Oxford University Press, New York, 2008.

Bradbury, Ray, *The Martian Chronicles*,1st Ed., Doubleday Publisher USA, 1950.

Breazeal, Cynthia, and Brooks, Rodney, *Robot Emotion: A Functional Perspective*, M.I.T Press, Cambridge, Massachusetts, 2004

Bunte, Jim, Hallman, Dave and Mueller, Heinz, *Vintage Toys: Robots and Space Toys*, Krause Publications, Wisconsin, USA, 1999.

Buscaglia, Leo, F., *Living, Loving and Learning*, Fawcett Books, USA,

1982.

Caudill, Maureen, *In Our Own Image: Building an Artificial Person*, Oxford University Press, New York, 1992.

Conroy, Pat, *A Lowcountry Heart: Reflections on a Writing Life*, Nan A.Talese/Doubleday, New York, 2016.

Dick, Philip K., *Do Androids dream of Electric Sheep?* Doubleday Publishers, New York USA, 1968.

Doyle, James F. (Ed.), *Educational Judgements: Papers in the Philosophy of Education*, Routledge and Kegan Paul, London and Boston, 1973.

Eliot, Thomas Stearns, *Poems 1909-1925*, 1st Ed., Faber and Faber, London, 23 November, 1925.

Eliot, Thomas Stearns, *A Choice of Kipling's Verse Made by T.S. Eliot with an Essay on Rudyard Kipling*, 1st Edition, Charles Scribner's Sons, New York, 1943.

Emchowicz, Antoni, and Nunneley,Paul, *Future Toys: robots,astronauts,spaceships, ray guns*, New Cavendish Books, Great Britain, 1999.

Ernst & Young LLP (UK), Global Tax Alert: *European Parliament calls for legislative framework governing creation, use and taxation of robots and artificial intelligence*, London, February 2017.

Fiorito,Piero, *Modelling Review Magazine (Rassegna di Modellismo)*, Italy, 1957.

Gibson, William, *Neuromancer*, Gollancz – The Orion Publishing Group, London, Great Britain, 1984.

Graetz, Georg and Michaels, Guy, *Robots at Work*, Discussion Paper No 1335, Centre for Economic Performance, London, March 2015.

Gunn, James, *Isaac Asimov: The Foundations of Science Fiction*, revised edition, The Scarecrow Press, Maryland and London, 1996.

Habyarimana, Bangambiki, *Pearls of Eternity:Musings of a Life Loving*

Soul, Smashwords ebook, 2016.

Herath, Damith, Kroos, Christian and Stelarc (Eds.) *Robots and Art: Exploring an Unlikely Symbiosis*, Springer Singapore, 2016.

Institute of Electrical and Electronics Engineers, *Proceedings of 2013 IEEE/RSJ International Conference on Intelligent Robots and Systems (IROS2013)*, 3-7 November 2013, Tokyo, Japan, Curran Associates, April 2014.

Japan External Trade Organisation, *Industrial Robots, Now In Japan Series, No.32/1981*, JETRO, Tokyo, 1981.

Kurzweil, Ray, *The Age of Spiritual Machines When Computers Exceed Human Intelligence*, Viking Press, New York, 1999.

Leebon, Timothy, *Predator: Incursion (The Rage War; Book 1)*, Titan Books, London, 25 September 2015.

Legal Affairs Committee, European Parliament, Draft Report 2015/2103 (INL): *Motion For A European Parliament Resolution –Civil Law Rules on Robotics*, 12 January 2017.

Levin, Ira., *The Stepford Wives*, 1st Ed., Random House, New York, 1972.

Liddell, Henry George, Scott, Robert, Jones, Henry Stuart, and McKenzie, Roderick, *A Greek and English Lexicon*, Clarendon Press, Oxford, 1940.

Longfellow, Henry Wadsworth, *In The Harbor, Ultima Thule. – Part II.*, 1st Ed., Houghton, Mifflin and Company, Boston, 1882.

Lovecraft, H.P., *Supernatural Horror in Literature*, The Recluse Magazine, Recluse Press, Massachusetts, USA, 1927.

Lucas, George, *Star Wars: Episode IV- A New Hope*, 20th Century Fox, USA, 1977.

MacDorman, Karl F., and Ishiguro, Hiroshi, 'Towards Social Mechanisms of Android Science', in Social Behaviour and Communication in Biological and Artificial Systems, *Interaction Studies Journal*, Volume 7, Issue

2, 2006.

Marshall, Percival (Ed.). *The Model Engineer and Light Machinery Review 59*, No.1429, 20 September 1928.

Matronic, Ana, *Robot Universe: Legendary Automatons And Androids From The Ancient World To The Distant Future*, Sterling Publishing, New York, 2015.

McConnell, Martin, *Viral Spark*, 1st Ed., New Land Press, 2016.

McKerrow, Phillip, *Introduction to Robotics*, Addison-Wesley Publishers, Boston, Massachusetts, USA, 1991.

Melzer, Patricia, *Alien Constructions: Science Fiction and Feminist Thought*, University of Texas Press, Austin, 2016.

Nadel, Laurie, *Dr. Laurie Nadel's Sixth Sense: Unlocking Your Ultimate Mind Power*, American Society of Journalists and Authors Press, New England, USA, 2006.

Nelson, Victoria, *The Secret Life of Puppets*, Harvard University Press, Cambridge, Massachusetts, 2001.

Oppenheim, Maya, *The Independent* newspaper, Independent Publications Limited, United Kingdom, 3 August 2017.

Picard, Rosalind, *Affective Computing*, M.I.T Media Laboratory Perceptual Computing Section Technical Report No. 321, M.I.T Press, Cambridge, Massachusetts,1997.

Radio Control Models and Electronics Magazine, MyTimeMedia, Kent, United Kingdom, Issue September 1960.

Rawson, Hugh, and Miner, Margaret, *The Oxford Dictionary of American Quotations*, 2nd Ed., Oxford University Press, USA, 2006.

Roberts, Dan, *Famous Robots and Cyborgs: An Encyclopedia of Robots from TV, Film, Literature, Comics, Toys and More*, Skyhorse Publishing, New York, 2014.

Ronson, Jon, *Lost At Sea: The Jon Ronson Mysteries*, Picador, United Kingdom, 2012.

Rosenberg, Jerry, M., *Dictionary of Artificial Intelligence and Robotics*, John Wiley & Sons Inc., New York, 1996.

Rosheim, Mark, *Robot Evolution: The Development of Anthrobotics*, John Wiley & Sons Inc, Canada, 1994.

Royakkers, Lambèrt and Van Est, Rinie, *Just Ordinary Robots: Automation from Love to War*, CRC Press, Boca Raton, Florida, 2016.

Russell, Ben (Ed.), *Robots: The 500-year quest to make machine human*, Scala Arts & Heritage Publishers Ltd, London, 2017.

Sandburg, Carl, *The Complete Poems of Carl Sandburg*, Houghton, Mifflin, Harcourt, New York, 2003.

Scientific American Journal, Vol 200, Issue 6, June 1959.

Severing, Kirsten (Ed.), Advanced Science Vol 2, Issue 10, WILEY-VCH Verlag GmbH & Co. KGaA, Weinheim, Germany, 15 October 2015.

Shalako, Louis Bertrand, *Twisted Tongue Magazine*, Issue #16, Claire Nixon Publisher, United Kingdom, June, 2010.

Siciliano, Bruno and Khatib, Oussama, (Eds.), *Springer Handbook of Robotics, Springer Science & Business Media*, Berlin, Heidelberg, 2008.

Space Family Robinson, *Gold Key Comics*, Western Publishing Company, Wisconsin, USA, December 1962 – October 1969.

Telotte, Jay P., *Robot Ecology and the Science Fiction Film*, Taylor and Francis, New York, 2016.

The Seattle Times, Issue 12, March 1993.

Truitt, Elly R., *Medieval Robots: Mechanism, Magic, Nature and Art*, University of Pennsylvania Press, Philadelphia, 2015.

Van Wagoner, Phillip and Smith, Spencer (Illustrator), *Could You Hug A Cactus?* Scribble Fiction, USA, 2016.

Vicente, Jose Manuel Ferrández et al.(Eds.), *Natural and Artificial Models in Computation and Biology, 5th International Work-Conference on the Interplay Between Natural and Artificial Computation (IWINAC 2013), Proceedings*, Part 1, June 10-14, 2013, Spain, Springer Science & Business Media, Heidelberg, Germany, Publishing, 2013.

Von Neumann, John, *Theory of Self-Reproducing Automata*, Arthur W. Burks (Ed.), University of Illinois Press, Urbana and London, 1966.

Weiner, Norbet, *Cybernetics*, or *Control and Communication in the Animal and the Machine*, 1st Ed., John Wiley & Sons, USA, 1948.

Wilson, Daniel, H., *How To Survive a Robot Uprising: Tips on Defending Yourself Against the Coming Rebellion*, 1st Ed., Bloomsbury Publishing, London, 2005.

Wyss, Johann David, *The Swiss Family Robinson*, Johann Rudolph Wyss publisher, Switzerland, 1812.

About the Author

The Robot Within Us All
(Frankonia, Japan, circa 1960s,
Image Credit: Shutterstock)

Simon King is an emerging Australian author who has already published three books: Crocodiles and Cocktails (Hesperian Press, 2017), Witchcraft, Whispers, Shadows and Strange Sights, and Marbles, Marella Jubes and Milk Bottles (Conscious Care Publishing, 2017). Each book encompasses different aspects of his life's journey, and engages many interesting historical and contemporary elements relating to these times.

Simon's fourth book adopts an unusual view of the challenging subject of robotics and artificial intelligence. As an avid sci-fi fan, he is able to embrace the fictional side to automata, robots, and androids through the media of print, television and movies. The technological advancements in robotics and cybernetics that are actually underway worldwide provide further elaboration on current research progress.

The merging of science fiction of the past with the many escalating scientific advancements in the 21st century have provided this author with quite a popular and enlightening topic for this book.